Advances in Social Science Research Using R

For other titles in this series, go to
www.springer.com/series/694

H.D. Vinod
Editor

Advances in Social Science Research Using R

Springer

Editor
Hrishikesh D. Vinod, Ph. D.
Department of Economics
Fordham University
441 E. Fordham Road
Bronx NY 10458
USA

ISBN 978-1-4419-1763-8 e-ISBN 978-1-4419-1764-5
DOI 10.1007/978-1-4419-1764-5
Springer New York Dordrecht Heidelberg London

Library of Congress Control Number: 2009942719

Printed on acid-free paper

Springer is part of Springer Science+Business Media (www.springer.com)

To My Daughter Rita,
Son-in-Law Dale Knause,
and My Grandchildren:
Devin Vinod Knause and Troy Vinod Knause

Preface

Quantitative social science research has been expanding due to the availability of computers and data over the past few decades. Yet the textbooks and supplements for researchers do not adequately highlight the revolution created by the R software [2] and graphics system. R is fast becoming the *lingua franca* of quantitative research with some 2000 free specialized packages, where the latest versions can be downloaded in seconds. Many packages such as "car" [1] developed by social scientists are popular among all scientists.

An early 2009 article [3] in the *New York Times* notes that statisticians, engineers and scientists without computer programming skills find R "easy to use." A common language R can readily promote deeper mutual respect and understanding of unique problems facing quantitative work in various social sciences. Often the solutions developed in one field can be extended and used in many fields. This book promotes just such exchange of ideas across many social sciences. Since Springer has played a leadership role in promoting R, we are fortunate to have Springer publish this book.

A *Conference on Quantitative Social Science Research Using R* was held in New York City at the Lincoln Center campus of Fordham University, June 18–19, 2009. This book contains selected papers presented at the conference, representing the "Proceedings" of the conference.

The conference offered an opportunity for enthusiastic users of R to discuss their research ideas and some policy applications in social sciences, to learn, meet and mingle. Speakers included authors of books and/or packages for R, published researchers and editors of important journals in statistics and social sciences. The keynote speaker was Roger Koenker, Professor of Economics, University of Illinois and there was a distinguished panel of invited speakers: Prof. Andrew Gelman, Director of the Applied Statistics Center at Columbia University; Prof. Kosuke Imai, Department of Politics at Princeton University; Prof. Keith A. Markus, Psychology Department at John Jay College of Criminal Justice; Prof. B. D. McCullough, Department of Decision Sciences, Drexel University and Prof. Achim Zeileis, Department of Statistics and Mathematics, WU Wirtschaftsuniversität Wien (Vienna).

We had a truly interdisciplinary conference organizing committee of Fordham professors. Co-chair: Frank Hsu (Computer and Information Sciences) and H. D. Vinod (Economics). Jose Aleman, treasurer (Political Science). Remaining Members (alphabetical order) were: Elena Filatova (Computer and Information Sciences), Se-Kang Kim (Psychology), Erick W. Rengifo (Economics), Emily Rosenbaum (Sociology), Matthew Weinshenker (Sociology), and Tiffany Yip (Psychology).

Fordham University was the main sponsor of the conference providing all the facilities, guidance and staff help. It would not have been possible without the active support from the deans of the Graduate School of Arts and Sciences (GSAS, Dean: Dr. Nancy Busch) Fordham College at Rose Hill (FCRH, Dean: Dr. Brennan O'Donnell) and College of Business Administration (CBA, Dean: Dr. Donna Rapaccioli). Another active supporter, Dr. Robert Himmelberg, Dean of Arts & Sciences Faculty, inaugurated the conference. Cosponsors were the Society of Indian Academics in America (SIAA) and Global Association of Risk Professionals (GARP). The modest registration fee paid for the coffee breaks and a cocktail reception with snacks on the evening of Thursday, June 18, 2009.

Although seating was limited and a registration fee was charged, seventy-five people registered and attended the conference. A limited number of graduate and undergraduate students from various disciplines attended the conference. Fordham University faculty and deans encouraged their students to take advantage of this opportunity to learn R. Students could learn some innovative and practical research tools providing intellectual skills having potentially lifelong benefits. The conference provided graduate students and young faculty in all disciplines a great chance to see, talk to and seek advice from researchers using R. The conference program and details remain available at the conference website: `http://www.cis.fordham.edu/QR2009`.

Most attendees thought that the conference was a great success. The research work presented was of high quality covering a wide range of important topics along with important examples and R implementations. This book of "Proceedings" extends the benefits of attending the conference to the much wider audience of academics and professionals in various sciences around the world. For example, one of our cosponsors (GARP, mentioned above) encouraged risk professionals who model risk and uncertainty in financial markets and who work in the metropolitan New York area to attend the conference.

A Brief Review of Each Chapter

Since all chapters come with their own abstracts, one may wonder why I am including this section. Note that our conference was interdisciplinary and individual authors have generally focused their abstracts to be appealing to readers from their own specific disciplines. I am including this section here because I believe that chosen papers are of interest to a much wider range of readers from other fields and also to those interested in the R software per se. Consistent with an aim of the conference, the following summaries are intended to promote greater cross fertilization of ideas among all scientists.

1] The chapter by McCullough shows the importance of numerically accurate computing indicating the relevance of forward and backward errors and condition numbers in all sciences. He shows that R software is accurate using some classic examples. A general reader might be interested in McCullough's reasons for questioning the reliability of Donohue and Levitt's controversial paper. These authors speculated that a disproportionate number of aborted (unwanted) children would have grown up to become criminals and claimed to show that abortion reduces crime after a gap of 20 years.

2] Koenker's chapter should be of interest to anyone seeking state-of-the-art flexible functional forms for regression fitting. The "additive" models seek a clever compromise between parametric and nonparametric components. Some commonly made data "adjustments" are shown to be misleading and avoidable. The R package "quantreg" implementation for modeling childhood malnutrition in India also should be of interest to policymakers. For example, mothers who are employed and wealthier are found to have taller children.

3] Gelman's chapter is of practical interest to all of us who use R for graphics and his example of U.S. voting participation rates is of interest to all who care about democracy and responsible citizenry.

4] My chapter suggests new solutions to a rather old and common problem of efficient estimation despite autocorrelation and heteroscedasticity among regression errors. Since regression is a ubiquitous tool in all sciences, it should be of much wider interest.

5] Markus and Gu's chapter is of great interest in exploratory data analysis in any scientific field. Given any data set, one often needs to develop initial understanding of the nature of relationships between three-way continuous variables. It is enhanced by comparing their new "bubble plots" conveniently made available as an R script called "bp3way()" at the conference website `http://www.cis.fordham.edu/QR2009`.

6] Vinod, Hsu and Tian's chapter deals with the portfolio selection problem of interest to anyone with an investment portfolio. It is also of interest to computer scientists and data-mining experts since combinatorial fusion comes from those fields. The idea of mapping various related variables into a comparable set ready for combining them has applications in many social sciences. Social scientists doing exploratory data analysis where lots of variables are potentially relevant should find these R tools useful.

7] Foster and Kecojević's chapter is of wide interest because they extend analysis of covariance (ANCOVA), a very general statistical tool. After all, many scientists want to know whether certain factors have an effect on a continuous outcome variable, while removing the variance associated with some covariates. Another common problem addressed here occurs when observations with the big residuals need to be down weighted. The wonderful (highly sophisticated) R graphics and growth charts for Saudi children are also of interest in their own right.

8] Imai, Keele, Tingley and Yamamoto's chapter addresses an age-old scientific problem of assessing the direction and strength of causation among variables. It is remarkably powerful, because of flexibility and variety in the types of cause–effect relations admitted. The authors explain the use of their R package called "mediation." Their "job search" application has wider research interest providing room for extensions, since the unemployment problem can be viewed from the distinct viewpoint of several social sciences.

9] Haupt, Schnurbus and Tschernig's chapter focuses on the problem of the choice of functional form for an unknown, potentially nonlinear relationship. This discussion goes beyond [4], where I discuss the related problem of choosing production functions. This chapter describes how to use flexible nonparametric kernel regression models for model validation of parametric specifications including misspecification testing and prediction simulations. The "relax" package is shown to be useful for model visualization and validation.

10] Rindskopf's chapter shows how to use R to fit a multinomial model that is parallel to the usual multivariate analysis of variance (MANOVA). A wide variety of categorical data models can be fit using this program, including the usual test of independence, logit and loglinear models, logistic regression, quasi-independence models, models with equality restrictions on parameters, and a variety of nonstandard loglinear models. An advantage of this program over some loglinear models is the clear separation of independent (predictor) and dependent (outcome) variables. Examples are presented from psychology, sociology, political science and medicine.

11] Neath's chapter dealing with location of hazardous waste sites also deals with a topic of wide general interest in policy circles. More important, Bayesian analysis of posterior distributions used to be less graphical and less intuitive before the recent availability of R packages. Neath shows how to use "WinBUGS" and "R2WinBUGS" interfaces in the context of an example, providing a template for applications in other fields.

12] Numatsi and Rengifo's chapter dealing with financial market volatility is of interest to anyone with an investment portfolio subject to financial crises. It studies time series models using FTSE 100 daily returns. Many fields have time series with persistent discrete jumps, misspecification and other issues which are solved by the authors' R software.

My short descriptions of twelve chapters explain why I am enthusiastic about including them in this book. They are my possibly inadequate sales pitches encouraging all readers to study the original chapters in detail. The authors of papers presented at the conference were given a short deadline of a couple of months for submission of their papers as a Latex document suitable for Springer publications. We are including only the ones who met the deadline and whose referees recommended their publication.

The original developers of the S computer language at Bell Labs (on which R is based) had insisted that the assignment operator must have two stroke "<-". When I worked at Bell Labs, I could not convince them to permit the use of one stroke "=" used by Fortran and other languages as a more convenient option for lazy typists. I am happy to see that unlike S, the R team does indeed permit the "=" symbol for assignment. Similarly, R gives me the option to start my R session with the command:

```
options(prompt = ";", continue = "      ")
```

This has the advantage that lazy typists like me can directly copy and paste selected lines of the output from R into my R input command stream. Without the 'options' command, the current prompt ">" is confused by R with "greater than," and the current continuation symbol "+" is confused with addition.

We enthusiastic users of R are grateful to the R Foundation for Statistical Computing, Vienna, Austria, [2] for continually maintaining and improving R. The R Foundation has nurtured a wonderfully flexible research tool, making it into a great intellectual resource for the benefit of humanity and deserves our thanks and financial support. Please see the separate section on 'Acknowledgments' listing names of those whose help was crucial in writing this book, including the referees.

Tenafly, New Jersey, October 2009 *Hrishikesh Vinod*

References

1. Fox, J.: car: Companion to Applied Regression. R package version 1.2-14 (2009). URL `http://CRAN.R-project.org/package=car`
2. R Development Core Team: R: A Language and Environment for Statistical Computing. R Foundation for Statistical Computing, Vienna, Austria (2009). URL `http://www.R-project.org`. ISBN 3-900051-07-0
3. Vance, A.: Data Analysts Captivated by R's Power. The New York Times **January 6** (2009). URL `http://www.nytimes.com/2009/01/07/technology/business-computing/07program.html`
4. Vinod, H.D.: Hands-on Intermediate Econometrics Using R: Templates for Extending Dozens of Practical Examples. World Scientific Publishers, Hackensack, NJ (2008). URL `http://www.worldscibooks.com/economics/6895.html`

Fig. 1 *Standing left to right:* Joachim Schnurbus, Rebecca Sela, Achim Zeileis, Tatjana Kecojević, David Rindskopf, Ronald Neath, Adjoa Numatsi and Rossen Trendafilov. *Sitting left to right:* B. D. McCullough, D. Frank Hsu, Roger Koenker, H. D. Vinod, Kosuke Imai and Keith Markus.

Acknowledgments

I am grateful to the Co-chair, Dr. Frank Hsu, of the organizing committee along with all the members whose names are mentioned in the Preface for their help with the conference. The following deans provided important encouragement: Nancy Busch, Brennan O'Donnell, Donna Rapaccioli and Robert Himmelberg.

My current students Rossen Trendafilov and Brandon Vick gave an "Introduction to R" talk which attracted the students. My former student Thethach Chuaprapaisilp helped greatly with typesetting and Latex issues. His help was crucial in incorporating the numerous suggestions by Springer's copy editor. Professor Elena Filatova helped tirelessly with the conference website and its updates. Our department secretary Giuseppina Cannariato was a big help throughout. Mr. John Kimmel, Executive Editor, Statistics at Springer, provided useful advice and encouragement over several months.

The quality of the papers depends critically on our distinguished and hard-working authors and referees, who provided timely critical comments and suggestions so the authors could improve the manuscripts. Some referees were also presenters of papers and some will remain anonymous. However, I do want to thank: William Anderson (Cornell University), Bruce L. Brown (Brigham Young University), Andreas Heinen (Universidad Carlos of Spain), Matthew S. Johnson (Columbia University), D. M. Lyons (Fordham University), and Philipp Sibbertsen (Leibniz Universität, Hannover, Germany), for their refereeing work.

Contents

List of Contributors

Peter Foster
School of Mathematics, University of Manchester, Manchester, UK
e-mail: peter.foster@manchester.ac.uk

Andrew Gelman
Department of Statistics and Department of Political Science, Columbia University, New York, NY 10027, USA e-mail: gelman@stat.columbia.edu

Wen Gu
Psychology Department, John Jay College of Criminal Justice, NY 10019, USA e-mail: wgu@gc.cuny.edu

Harry Haupt
Centre for Statistics,Department of Economics and Business Administration, Bielefeld University, 33501 Bielefeld, Germany
e-mail: hhaupt@wiwi.uni-bielefeld.de

D. Frank Hsu
Department of Computer & Information Science, Fordham University, New York, NY 10023, USA e-mail: hsu@cis.fordham.edu

Kosuke Imai
Department of Politics, Princeton University, Princeton, NJ 08544, USA
e-mail: kimai@princeton.edu

Tatjana Kecojević
Lancashire Business School, University of Central Lancashire, Preston, UK
e-mail: TKecojevic@uclan.ac.uk

Luke Keele
Department of Political Science, Ohio State University, Columbus, OH 43210, USA e-mail: keele.4@polisci.osu.edu

Roger Koenker
Department of Economics, University of Illinois, Champaign, IL 61820, USA
e-mail: rkoenker@uiuc.edu

Keith A. Markus
Psychology Department, John Jay College of Criminal Justice, New York, NY 10019, USA e-mail: kmarkus@aol.com

B. D. McCullough
Department of Economics, Drexel University, Department of Decision Sciences, LeBow College of Business, Drexel University, Philadelphia, PA 19104, USA e-mail: bdmccullough@drexel.edu

Ronald C. Neath
Department of Statistics and CIS, Baruch College, City University of New York, New York, NY 10010, USA e-mail: ronald.neath@baruch.cuny.edu

Adjoa Numatsi
Department of Economics, Fordham University, Bronx, NY 10458, USA
e-mail: numatsi@fordham.edu

Erick W. Rengifo
Department of Economics, Fordham University, Bronx, NY 10458, USA
e-mail: rengifomina@fordham.edu

David Rindskopf
Educational Psychology Program, CUNY Graduate Center, New York, NY 10016, USA e-mail: drindskopf@gc.cuny.edu

Joachim Schnurbus
Institute of Economics and Econometrics, University of Regensburg, 93053 Regensburg, Germany
e-mail: joachim.schnurbus@wiwi.uni-regensburg.de

Ye Tian
Department of Applied Mathematics and Computational Science, University of Pennsylvania, Philadelphia, PA 19104, USA
e-mail: ytian001@hotmail.com

Dustin Tingley,
Department of Politics, Princeton University, Princeton, NJ 08544, USA
e-mail: dtingley@princeton.edu

Rolf Tschernig
Institute of Economics and Econometrics, University of Regensburg, 93053 Regensburg, Germany e-mail: rolf.tschernig@wiwi.uni-regensburg.de

Hrishikesh Vinod
Fordham University, Bronx, NY 10458, USA e-mail: vinod@fordham.edu

Teppei Yamamoto
Department of Politics, Princeton University, Princeton, NJ 08544, USA
e-mail: tyamamot@princeton.edu

Chapter 1
Econometric Computing with "R"

B. D. McCullough

Abstract We show that the econometrics profession is in dire need of practitioners of econometric computing, and that "R" is the best choice for teaching econometric/statistical computing to researchers who are not numerical analysts. We give examples of econometric computing in R, and use "R" to revisit the classic papers by Longley and by Beaton, Rubin and Barone. We apply the methods of econometric computing to show that the empirical results of Donohue and Levitt's abortion paper are numerically unsound. This discussion should be of interest in other social sciences as well.

1.1 Introduction

Econometric computing is not simply *statistical computing* for econometrics; there is much overlap between the two disciplines, yet there are separate areas, too. Econometric computing is, to a large extent, the act of making econometric theory operational. This stands in sharp contrast to statistical computing, which is not at all defined by its users (e.g., biologist, physicist, or any of scores of disciplines). Statistics has not much to say about whether some particular empirical measurement might exhibit a unit root, while the econometric literature has a cottage industry dedicated to unit roots. These themes are explored in the recent book by Renfro [26].

Practitioners of statistical computing will not, in general, concern themselves with econometric issues such as cointegration or three-stage least squares, yet economists have need of these procedures and the accuracy and stability of these procedures should be addressed. This is the purview of econometric computing. There is a tendency by some to use statistical

B. D. McCullough
Department of Economics, Drexel University, Philadelphia, PA 19104, USA
e-mail: bdmccullough@drexel.edu

H.D. Vinod (ed.), *Advances in Social Science Research Using R*,
Lecture Notes in Statistics 196, DOI 10.1007/978-1-4419-1764-5_1,

(econometric) computing and computational statistics (econometrics) interchangeably, and this leads to confusion. To clarify, we quote from Zeileis [33, p. 2988]:

> [T]here is a broad spectrum of various possibilities of combining econometrics and computing: two terms which are sometimes used to denote different ends of this spectrum are (1) computational econometrics which is mainly about methods that require substantial computations (e.g., bootstrap or Monte Carlo methods), and (2) econometric computing which is about translating econometric ideas into software. Of course, both approaches are closely related and cannot be clearly separated[.]

Implicit in the above definition of econometric computing is that the software created be correct. More conventionally, statistical computing is numerical analysis for statistics, while computational statistics is the use of computationally intensive methods (e.g., simulation, bootstrap, MCMC, etc.) to conduct inference. We make the same distinction for econometric computing and computational econometrics.

Econometric computing has been even less prominent in econometrics than has statistical computing in statistics, to the detriment of the practice of economics. In general, economists know nothing of econometric computing. Evidence suggests that some developers of econometric software know little of the subject, too — witness how frequently two or more econometric software packages give different answers to the same problem. The solution to the problem is the development of a course in econometric computing to be taught to graduate students in economics. It is the contention of this paper that "R" is the only program capable of serving as a vehicle for teaching econometric computing.

Of course, if "R" is to be the solution, it must first be demonstrated that a problem exists. Hence, Section 1.2 contends that most users and some developers of econometric software are generally unaware of the principles of econometric computing. Continuing in this vein, Section 1.3 presents some examples to show that accurate computing matters. Section 1.4 discusses the reasons that "R" is the only viable language for teaching econometric computing. Section 1.5 reexamines Longley's [13] classic paper. Section 1.6 presents Beaton, Rubin and Barone's [2] reanalysis of Longley. Section 1.7 applies the lessons of Beaton, Rubin and Barone to the abortion paper by Donohue and Levitt [6] that was recently analyzed by Anderson and Wells [1] from a numerical perspective. Our results support those of Anderson and Wells: the Donohue and Levitt results are numerically unreliable.

1.2 The Economics Profession Needs Econometric Computing

There is substantial overlap between statistics and econometrics, yet there are econometric methods of which most economists are aware and of which most statisticians will never hear. Similarly for statistical computing and econometric computing. Econometrics cannot free-ride on statistical computing, because there are many econometric issues that practitioners of statistical computing will never address.

Consider, for example, Generalized Autoregressive Conditional Heteroscedasticity (GARCH), for which Rob Engle won the Nobel Memorial Prize in economics. Hundreds, if not thousands of GARCH articles had been written and published during a period when no two software packages gave the same answer to the same GARCH problem. Fiorentini, Calzolari and Panattoni [9] produced benchmark-quality code for solving the GARCH problem. McCullough and Renfro [21] demonstrated that different packages gave different answers to the same GARCH problem and used the Fiorentini et al. code to produce a GARCH benchmark. Software developers converged on this benchmark and subsequently many packages gave the same answer to the same GARCH problem, as demonstrated by Brooks et al. [4]. Another such econometric computing problem is *cointegration*, for which Clive Granger won the Nobel Memorial Prize in economics, the same year as Engle, and which is discussed in Sect. 1.3.

Experts in statistical computing will never address these econometric issues; only economists will. But before economists can address these econometric computing issues, they need to know the elements of statistical computing. In general, they lack this knowledge.

1.2.1 Most Users Do Not Know Econometric Computing

It stands to reason that researchers who write their own code are more likely to be acquainted with the fundamentals of computing than researchers who use canned procedures. Yet, when examining the code written by these researchers, it is easy to find examples of computing illiteracy. There are archives of code written by social science researchers in many languages: MATLAB, GAUSS, and even "R". Examining the code in these archives will reveal numerous instances of computing illiteracy, here we mention only three:

1. solving for the least squares regression estimator as b = inv(X'X)*X'y (see McCullough and Vinod [22, §2.5])
2. writing log(norm(x)) instead of lognorm(x) (see McCullough and Vinod [23, p. 880])

3. computing square roots as x**0.5 instead of sqrt(x) (see Monahan [24, p. 153])

Users have no idea that what they are doing is wrong with potentially disastrous consequences for their estimation results.

1.2.2 Some Developers Do Not Know Econometric Computing

Examining the track-record of software developers, it is hard to escape the conclusion that more than a few of them are unacquainted with the fundamentals of statistical computing:

1. using the calculator formula to compute the sample variance (see McCullough and Vinod, [22, §2.5])
2. not offering lognorm(x)
3. computing tails of distribution functions via complementation rather than computing them directly (see McCullough and Vinod [22, §6])
4. using the Yule–Walker equations to solve for autocorrelation coefficients (see McCullough [14] for an extended discussion)
5. solving maximum likelihood cointegration problems using the Cholesky decomposition (see Sect. 1.3 of this paper)

Numerous other examples of poorly programmed commercial econometric software are provided in McCullough [20]. As further evidence that one should not be surprised to find software developers unable to program correctly even trivial code (e.g., the "sample variance" and "correlation coefficient" examples noted above), we further note that the largest programming company in the world (Microsoft) was unable to implement the dozen lines of code that constitute the Wichmann–Hill random number generator — twice! See McCullough [18] for details and discussion.

Both the above lists could be extended dramatically, but the point is clear: economists typically know little about econometric computing, i.e., how to make sure that the computer is delivering an accurate answer.

1.2.3 Some Textbook Authors Do Not Know Econometric Computing

In the recently published fourth edition of *Essentials of Econometrics* by Gujarati and Porter [11], the authors use four software packages throughout the text, often using all four packages to solve the same problem. In Chapter 1, for the very first regression problem, the students are shown the following equation:

$$CLFPR = \beta_0 + \beta_1 CUNR + \beta_2 AHE82 + \varepsilon$$

based on 28 observations, 1980–2007 from the *Economic Report of the President, 2008*, with the data presented in Table 1.1 of the book.

Also presented in the book are "results" from EViews, Excel, Minitab and Stata, which are given below in Table 1.1, along with results from using R to estimate the equation.

Table 1.1 Gujarati/Porter results

package	$\hat{\beta}_0$	$\hat{\beta}_1$	$\hat{\beta}_2$
EViews/Excel	81.22673	0.638362	-1.444883
Minitab/Stata	81.286	0.63877	-1.4521
R (correct)	81.30500	0.63846	-1.45477

As far as the EViews/Excel and Minitab/Stata answers are concerned, if one pair of packages is correct, then the other pair must be incorrect: which pair is correct and which is incorrect? Gujarati/Porter do not say. There is no doubt that "R" gives the correct answer. There is no doubt that EViews, Excel, Minitab and Stata have accurate regression routines, at least accurate enough to handle this simple problem. Clearly, Gujarati/Porter managed to give the same incorrect input dataset to EViews and Excel, and managed to give a different incorrect input dataset to Minitab and Stata.[1]

What is shocking is not that Gujarati and Porter made a mistake, but that they make no remark in the text about the fact that four packages give two answers to one problem! Their silence on this point has two striking implications:

1. They calculated two different answers to the same simple linear regression problem and found nothing unusual in this — they apparently are unaware that statistical and econometric software has progressed in the forty-plus years since Longley (1967 — to be discussed in Sect. 1.5) to the point that a user can reasonably expect different packages to give the same answer to

[1] Thanks to Houston Stokes, developer of the B34S software package, for pointing out this problem, and for working with me to reverse engineer the mistakes made by Gujarati/Porter. The data apparently were taken from http://www.gpoaccess.gov/eop/tables08.htmlwhere the requisite .xls files are available, formatted to display 1 decimal for CLFPR and CUNR, 2 decimals for AHE82 — this is what Gujarati/Porter show in their text. The spreadsheet, however, contains the data to several decimals in some cases, and using the actual data instead of the displayed data reproduces the EViews/Excel results. Apparently Gujarati/Porter loaded the spreadsheet into the software without checking to make sure that they loaded the data they actually wanted. To obtain the Minitab/Stata results requires yet another mistake. If the displayed data are used and the last observation for CUNR is changed from 66.0 to 66.2, then the Minitab/Stata results can be obtained.

the same linear regression problem. We have made abundantly clear that this is not true for most other procedures, but it is true for OLS.
2. They think *nothing* of presenting two different sets of solutions to the same problem from four different packages to students without remarking on it! They do not warn the student that this is usual — they can expect different packages to give different answers to the same problem; neither do they warn that this is unusual (for linear regression, at least). They simply present these disparate results and allow the student — if he notices — to form his own conclusion about the accuracy (or lack thereof) of econometric and statistical software.

It is not just Gujarati and Porter, but most econometrics text authors. For example, when treating nonlinear estimation, they simply tell the reader to use a software package to solve the problem, never warning the unsuspecting reader that different packages give different answers to the same problem, that some packages have horrible nonlinear solvers, etc. A notable exception is the recent text by Vinod [30].

One would hope that the authors of an econometrics text would have at least some passing familiarity with the accuracy (or lack thereof) of the software they use, but such is not the case. (Or, if they do possess this knowledge, they do not think it worth passing on to the student!) Crucially, they do not inform the student that computing is not error free, and if the economics student does not learn it from the authors of econometrics texts, from where is he to learn it? Regrettably, nowhere.

Despite the stance of econometric textbook authors, accuracy is important and it is not safe to assume that software is accurate (McCullough [16]).

1.3 Econometric Computing Is Important

In this section we present a few examples of computing gone bad; for more examples and detailed discussion, see McCullough [19] and the references therein.

- Different packages give different answers for the same ARMA problem [25], while McCullough [19] eventually provided benchmarks for conditional and unconditional least squares ARMA estimation — though maximum likelihood ARMA estimation is still an open question.
- Different packages give different answers to the same vector autoregression problem [20].
- Different packages give different answers to the same GARCH problem [21], [4].
- Different packages give different answers to the same multivariate GARCH problem [5].
- Different packages give different answers to the same FIML problem [27].

- Different packages give different answers to the same 3SLS problem [34].
- Different packages give different answers to the same logistic regression proble, and even find "solutions" when the solution does not exist [28].
- Packages calculate correlation coefficients greater than unity [17].
- Packages incorrectly calculate the sample variance [17], [15].
- Nonlinear solvers giving incorrect answers to nonlinear least squares problems [15], [29], [32].

Researchers can take note of these problems, but are unable to comprehend the problems without some knowledge of econometric computing. Here is an example.

Cointegration is an important concept in econometrics. For those unfamiliar with the idea, consider two random walk variables, x and y. Each is "integrated of order one", i.e., its first difference is a stationary sequence with finite variance. Each random walk series has (theoretically) infinite variance and wanders without bound as time increases. Suppose, however, that x and y cannot wander too far apart — their difference remains stationary with finite variance — then the variables are said to be "cointegrated". As an example, consider a spot price and a futures price for the same commodity. As financial variables in speculative markets, each is a random walk variable. Clearly, however, if the spot and futures prices diverge too much, then traders seeking profits will execute trades that bring the prices back together. A spot price and a futures price are cointegrated.

In cointegration analysis, systems of correlated, trending variables (whose design matrix will tend toward multicollinearity and hence ill-conditioning) are examined for cointegration using a method that requires solving a generalized eigenvalue problem:

$$|\lambda S_{11} - S_{10}S_{00}^{-1}S_{01}| = 0$$

In the econometrics literature (and in most econometrics packages that offer this procedure), the recommended way to solve this problem is to apply a Cholesky decomposition to S_{11} and reduce the problem to a standard eigenvalue problem.

The problem with this approach is that S_{11} can be very ill-conditioned, and the Cholesky decomposition will not then give a good answer; other methods such as the QR and the SVD are to be preferred. This idea is explored in detail in the excellent article by Doornik and O'Brien [7]. They generate a simple cointegrated system by

$$\mathbf{y_t} = (y_{1t}, y_{2t}); y_{2t} = y_{1t} + u_t 10^{-m}; u_t \sim N(0,1)$$

and then use various methods to solve the related generalized eigenvalue problem, with results given in Table 1.2 (inaccurate digits in bold). How can this problem possibly be explored in the context of econometric computing using

Table 1.2 Largest eigenvalue of system

Alg.	$m = 3$	$m = 5$	$m = 10$
Cholesky-1	0.35091940555	0.05949553867	failed
Cholesky-2	0.35091940557	0.13273076746	failed
QR-1	0.35091938503	0.35091938542	0.01335798065
QR-2	0.35091938503	0.35091938540	failed
SVD	0.35091938503	0.35091938494	0.00487801748

a standard econometrics software package that offers only generic, unspecified matrix inversion and eigenvalue calculation routines? It cannot!

1.4 "R" Is the Best Language for Teaching Econometric Computing

Econometric computing requires software that provides the numerically literate user the ability to "code" econometric techniques using high-quality components (subroutines); a generic matrix inversion routine does not satisfy this requirement. To teach econometric computing similarly requires a package that offers a wide range of relevant functionality. The traditional econometrics packages simply do not offer the necessary commands. A standard econometric computing task is to consider the differences between varying methods of solving for linear least squares coefficients; another standard task is to do the same for solving eigenvalue problems. Here is the complete description of the command to invert a matrix from EViews version 5:

```
Syntax:   @inverse(M)
Argument: square matrix or sym, m
Return:   matrix or sym

Returns the inverse of a square matrix object of sym.
The inverse has the property that the product of the
source matrix and its inverse is the identity matrix.
The inverse of a matrix returns a matrix, while the
inverse of a sym returns a sym.  Note that inverting
a sym is much faster than inverting a matrix.

Examples:
  matrix m2 = @inverse(m1)
  sym s2 = @inverse(s1)
  sym s3 = @inverse(@implode(m2))
```

For purposes of econometric computing, this command is worthless. The documentation provides no information as to the method used for inverting the matrix. The EViews command for solving an eigensystem is equally worthless for econometric computing. EViews is not unique in this regard; other econometrics packages (e.g., SHAZAM and RATS) have similarly-documented commands. One would hope that a software developer would appreciate the importance of letting a user know whether matrix inversion is performed by QR or Cholesky, but such is the state of the economics profession and econometric software.

The package MATLAB of course does not suffer from the above syndrome, but has impediments of its own. First, MATLAB (or other commercial systems) are not free. Students already have to acquire a commercial econometrics package. The acquisition of a second package is onerous. Open-source MATLAB clones such as Octave and SciLab are free, but this raises the second (and more important) impediment: popularity. Neither MATLAB nor its clones is sufficiently popular with statisticians that it is widely used for statistical computing — "R" is, and econometric computing gets a free ride from this. For example, the superb book by Monahan [24], *Numerical Methods of Statistics*; originally was released with FORTRAN files to illustrate examples. Monahan later added files of "R" code to achieve the same ends as the FORTRAN files. Indeed, Monahan's book would be an excellent choice for a first course in econometric (or statistical) computing, and "R" is much easier to learn and to use for this purpose than FORTRAN.

A standard exercise in statistical computing is inverting the Hilbert matrix. The Hilbert Matrix is a square matrix given by

$$H_{ij} = \frac{1}{i+j-1}$$

so for small cases, $n = 3$ and $n = 4$, it looks like this:

$$\begin{bmatrix} 1 & \frac{1}{2} & \frac{1}{3} \\ \frac{1}{2} & \frac{1}{3} & \frac{1}{4} \\ \frac{1}{3} & \frac{1}{4} & \frac{1}{5} \end{bmatrix} \begin{bmatrix} 1 & \frac{1}{2} & \frac{1}{3} & \frac{1}{4} \\ \frac{1}{2} & \frac{1}{3} & \frac{1}{4} & \frac{1}{5} \\ \frac{1}{3} & \frac{1}{4} & \frac{1}{5} & \frac{1}{6} \\ \frac{1}{4} & \frac{1}{5} & \frac{1}{6} & \frac{1}{7} \end{bmatrix}$$

The inverse of the Hilbert matrix has a closed form and so can be computed precisely:

$$H_{ij}^{-1} = (-1)^{i+j}(i+j-1)\binom{n+i-1}{n-j}\binom{n+j-1}{n-i}\binom{i+j-2}{i-1}^2$$

Yet, traditional numerical matrix inversion routines quickly run into trouble as n increases because the condition number of the Hilbert matrix grows like $\exp(3.5n)$. Consequently, relatively small Hilbert matrices can be used to demonstrate various aspects of ill-conditioned matrices, as follows.

Recall that $HH^{-1} = I_n$. The standard statistical computing problem then is to compute HH^{-1} and examine its main diagonal for various values of n and various methods of computing the inverse. For $n = 12$, and the two methods "Cholesky" and "SVD", the results from "R" are presented in Table 1.3.

Table 1.3 Main diagonal of HH^{-1} for Hilbert matrix of order 12

	Cholesky	SVD
1	1.0000002	1.0000000
2	0.9999807	1.0000012
3	1.0005020	0.9999702
4	0.9944312	1.0002072
5	1.0383301	0.9990234
6	0.8465960	0.9966422
7	1.3649726	0.9946361
8	0.3687544	0.9902124
9	1.6553078	1.0111723
10	0.5280704	1.0308642
11	1.1820953	1.0003664
12	0.9688452	1.0012815

How is this statistical computing exercise to be undertaken by an econometrics package that offers only an unspecified, generic method of calculating the inverse of a square matrix? The typical econometrics package is not up to the task.

On a related note, a very important quantity in statistical computing is the condition number of a matrix. One common way of computing this quantity is to take the ratio of the largest to the smallest eigenvalues. Yet, this requires an accurate eigenvalue routine. Is the unspecified, generic "eigenvalue" command typically found in an econometrics package up to the task? We do not know, because the developers typically present the routine as black box — and black boxes are not to be trusted. Again, we find that typical econometrics packages are unable to perform routine operations that are critical to statistical computing.

1.5 The Longley Data and Econometric Computing

In 1967, Longley [13] computed by hand the solution to a regression problem. Below are the data:

Longley's Data

y	x1	x2	x3	x4	x5	x6
60323	83.0	234289	2356	1590	107608	1947
61122	88.5	259426	2325	1456	108632	1948

60171	88.2	258054	3682	1616	109773	1949
61187	89.5	284599	3351	1650	110929	1950
63221	96.2	328975	2099	3099	112075	1951
63639	98.1	346999	1932	3594	113270	1952
64989	99.0	365385	1870	3547	115094	1953
63761	100.0	363112	3578	3350	116219	1954
66019	101.2	397469	2904	3048	117388	1955
67857	104.6	419180	2822	2857	118734	1956
68169	108.4	442769	2936	2798	120445	1957
66513	110.8	444546	4681	2637	121950	1958
68655	112.6	482704	3813	2552	123366	1959
69564	114.2	502601	3931	2514	125368	1960
69331	115.7	518173	4806	2572	127852	1961
70551	116.9	554894	4007	2827	130081	1962

X1 = GNP deflator, X2 = GNP, X3 = unemployment, X4 = size of armed forces, X5 = noninstitutional population aged 14 and over, X6 = time

The linear model considered by Longley was

$$Y = c + \beta_1 X1 + \beta_2 X2 + \beta_3 X3 + \beta_4 X4 + \beta_5 X5 + \beta_6 X6 + \varepsilon$$

Longley gave the problem to several regression packages. Most of them failed miserably: "With identical inputs, all except four programs produced outputs which differed from each other in every digit" [13, p. 822]. In Table 1.4 are Longley's hand-calculated coefficients and the output from a representative program that Longley tested, run in single-precision (Longley often gave the problem to a package twice: once without the means subtracted, and again with the means subtracted).

Table 1.4 Regression results by Longley and by a computer

coeff	longley	IBM 7074 uncentered	IBM 7074 centered
c	-3482258.63 30	-269126.40	+0.00876
b1	+15.061872271	-36.81780	+15.18095
b2	-0.035819179	+0.059053	-0.03588
b3	-2.02022980 3	-0.59308	-0.202104
b4	-1.033226867	-0.60657	-1.03339
b5	-0.05110410 5	-0.34354	-0.05072
b6	+1829.1514646 1	+183.73361	+1829.6936

These results validate a pair of statistical computing folk theorems. First, extremely ill-conditioned data can result in completely inaccurate results. The "condition number" (κ) of the Longley data is about $15000 (\approx 10^4)$, which indicates severe ill-conditioning. It is of little surprise that the program returned completely inaccurate results for the uncentered data. The algorithm could return somewhat accurate results for the centered data, and this validates the second folk theorem: if the data in y and X are accurate to about s digits and $\kappa(X) \approx 10^t$, then the *computed solution* is accurate to about $s-t$ digits. In single precision with about 7 digits to work with, we can expect that the coefficients can be calculated accurately to about 3 digits and that is what we see.

The history of the Longley paper in statistical computing has emphasized the effect of ill-conditioning on accurately computing coefficients. What has been lost is the message that ill-conditioning has dramatic implications for model adequacy even when the coefficients are accurately computed. This theme was explored by Beaton, Rubin and Barone [2] ten years later, but the paper did not received much attention in the literature. We attempt to make up for this oversight in the next section.

1.6 Beaton, Rubin and Barone Revisit Longley

We know from statistical computing that the condition number has import for the accuracy of calculations. It also has import whether the data are up to the task of estimating the specified model. Consider the Longley regression and the Longley data. Imagine perturbing the data within the limits to which the data were rounded. As an example, change the first observation on X1 from 83.0 to 83.0 + u[-0.499,0.499] where u[a,b] is a random uniform draw from [a,b]. Truthfully, the first observation on X1 is not known precisely to be 83.0, and could have been any number from 82.501 to 83.499. Changing the data beyond the limits to which they are reported should have no substantial effect on the estimated coefficients, provided that the data are sufficiently accurate, the algorithm is stable, and the model is approximately correct. Imagine perturbing all the observations this way and re-estimating the model. Do this 1000 times, getting 1000 estimates of each coefficient and make histograms of the estimates. Beaton, Rubin and Barone [2] did this, and their histograms are remarkably similar to the ones displayed in Fig. 1.1, where the vertical dashed line indicates the initial OLS coefficient. We call these "BRB plots".

What has happened? Why do such seemingly small differences in the input data have such large effects on the coefficients? The reason is that the model is unstable — the data are not up to the task of estimating this model. It is important to remember that the Longley data have only 16 observations — if the sample size were larger, the effects of perturbing the data beyond the last reported value would be much mitigated.

Due to finite precision (see McCullough and Vinod [22, §2]), the data in the computer do not match the real-world data, e.g., 0.1 is actually stored in a PC as 0.09999999403953. When such small differences are accumulated over the millions of operations necessary to compute regression coefficients,

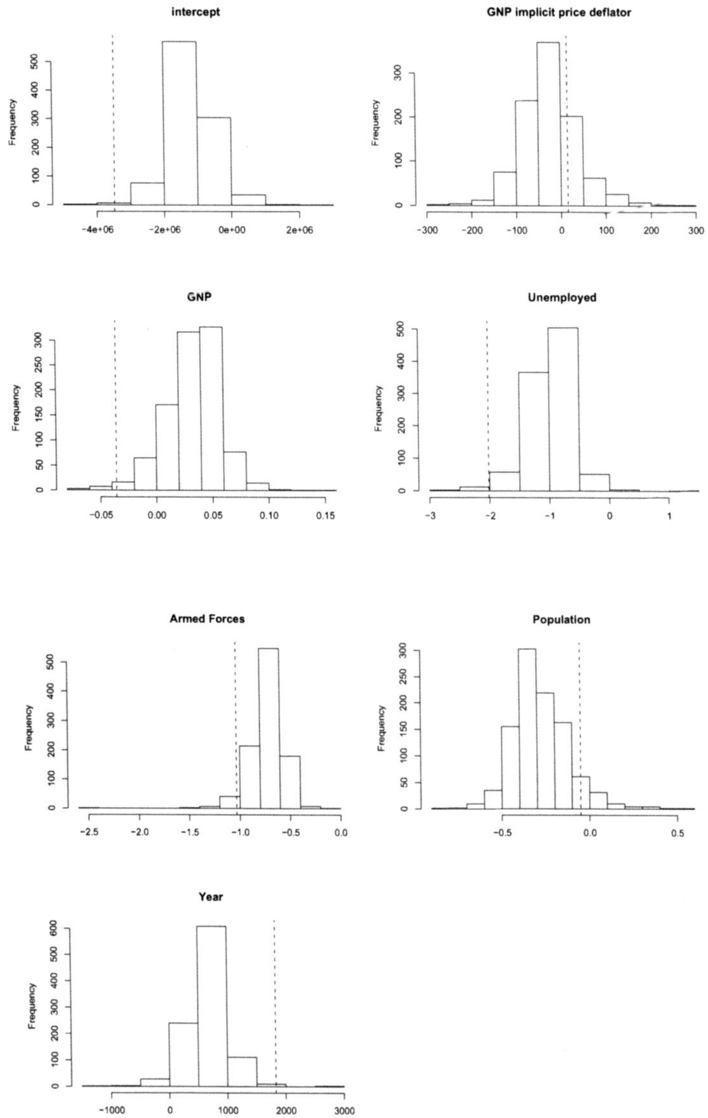

Fig. 1.1 BRB plots of the Longley coefficients.

large differences in the coefficients can result. More specifically, the computed solution $\hat{\beta}$ is not a precise solution to the original problem $y = X\beta + \varepsilon$ but instead is a precise solution to a nearby problem expressed in terms of y^* and X^* where $y^* = y + e^*$ and $X^* = X + E^*$.

The question is whether this "nearby" problem is really close to the problem of interest. In the present case, it is not. To see why not, it is useful to consider two fundamental ideas from numerical analysis: the *forward error* and the *backward error*. In forward error analysis, one attempts to find bounds for the solution of the estimated coefficients. Let $\hat{\beta}^*$ be the actual solution to the original problem. Forward error analysis is concerned with determining how close $\hat{\beta}$ is to $\hat{\beta}^*$. In backward error analysis, one attempts to find a perturbed set of data so that the computed solution is the exact solution for these perturbed data. These ideas are codified in what may well be the most important equation in all of numerical analysis:

$$\text{forward error} \leq \text{condition number} \cdot \text{backward error}$$

The graphs are evidence of a large backward error. Combined with the large condition number, we conclude that the computed solution is not assured to be close to the exact solution. See Hammarling [12] for a brief introduction to the concept of a "computed solution".

Gentle [10, p. 346] observes that ill-conditioning usually means that "either the approach or the model is wrong." With such a thought in mind, Beaton, Rubin and Barone adopted a simpler model

$$Y = c + \beta_1 X1 + \beta_3 X3 + \beta_4 X4 + \varepsilon$$

solely on the basis that X3 and X4 are relatively uncorrelated with each other and the other explanatory variables, and X1 is highly correlated with X2, X5 and X6. The BRB plots for this model are given in Fig. 1.2, and show a marked improvement. With the exception of the intercept term, the initial OLS estimates are nearer the center of the histograms. This strongly suggests that the reduced model is more consonant with the data than the full data. Note, too, how the values of the coefficients have changed dramatically when the model was respecified. Thus, the initial BRB plots do *not* tell us anything about the likely values of the coefficients — they only tell us that the exact model is far from the estimated model. Vinod [31] advocates the use of a "perturbation" index for a similar assessment of model quality.

1.7 An Example: Donohue/Levitt's Abortion Paper

With the above in mind, we turn our attention to Donohue and Levitt's (DL, [6]) paper on abortion, which argued that increases in abortion lead to a decrease in the crime rate 20 years later because a disproportionate percentage of the aborted children would have grown up to become criminals. The data and associated Stata code can be found at Donohue's website. Consider DL's Table IV, which shows results for three regressions with three dependent variables: log of violent crime per capita, log property crime per capita, and log of murder per capita. Each regression has one unique independent

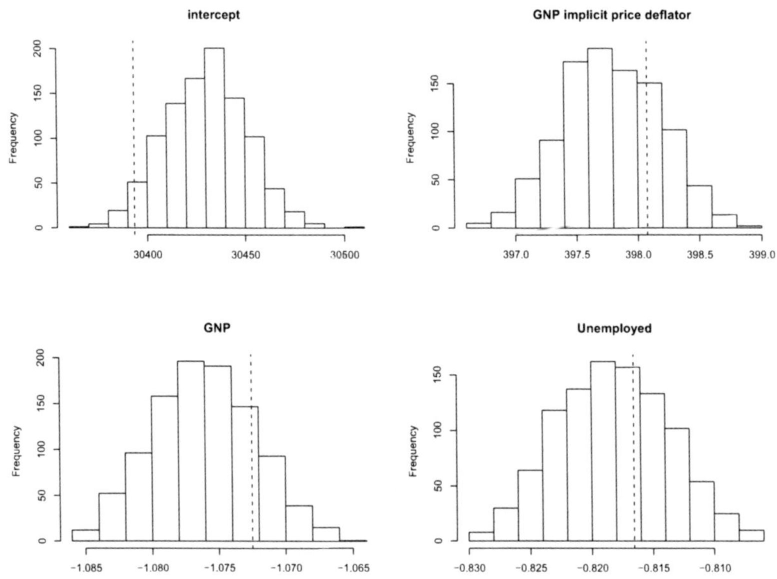

Fig. 1.2 BRB plots of the Longley coefficients, reduced model.

variable (effective abortion rate for violent crime, effective abortion rate for property crime, and effective abortion rate for murder, respectively) and 71 common independent variables divided between seven continuous covariates, one dummy variable, 50 state dummies (including Washington, DC), 12 year dummies, and one constant. Each regression has a 663×72 design matrix (the three design matrices differ only in one column). DL use Stata to run their linear regressions with a serial correlation correction for panel data. In attempting to replicate DL, Foote and Goetz (FG, [8]) discovered a coding error whereby DL neglected to include state-year effects, and offered other regressions to dispute DL's conclusions.

Both DL and FG were reanalyzed by Anderson and Wells (AW, [1]) who approached the problems from a numerical perspective. The first part of the AW paper provides the necessary technical background in numerical analysis. In the second part, among other things, AW show that

- The condition number of the 663×72 design matrix (when the observations are weighted by state population) is $\kappa = 1{,}329{,}930$, which is very large. Based on this, they conclude, "There cannot be any accuracy in the computed solution for a design matrix of less than six significant digits. In other words, there is not enough information in the data to meaningfully estimate the regression coefficients $\hat{\beta}$.

- The bound on the relative error of the estimated coefficents for DL's regression in their Table 5 is 530E+9. They conclude, "This bound is too high to have confidence in the computed solution."
- The bound on the relative error of the estimated coefficients for Table 1 of FG is 30E+6, which is very large. Again, AW conclude, "This upper bound is too high to have confidence in the computed solution."

In general, AW show that both DL and FG created models that were too demanding for the data.

Our analysis here follows a related but different approach to show that the DL model was too demanding for the data. For ease of exposition, our analysis will be limited to simple linear regression. Additionally, DL weighted the observations by population, to keep things simple we do not. AW adduce much evidence to indicate that the exact model is far from the estimated model e.g., condition number, variance inflation factors, etc.). We add a visual diagnostic as additional evidence on this point: we employ BRB plots. Third, because of the sample size there is no chance that perturbing beyond the last digit will have any effect, so we perturb based on the accuracy of the data. To this end, we first consider DL's data.

DL has a "data appendix" that is not much use in locating the actual data used by DL. For example, the entire entry for the Income variable is: "Per capita state personal income, converted to 1997 dollars using the Consumer Price Index, from Bureau of the Census, *United States Statistical Abstract* [annual]." Since data are revised, it is important to know the year of publication of the *United States Statistical Abstract*; since it will sometimes have different tables of per capita income under different definitions, it is important to know the specific table from which the data were taken. Similarly for the Consumer Price Index. DL made it hard for a replicator to locate the sources of their data; we do not feel bound to waste resources tracking them down. In general, we could not readily find data series that match DL's data precisely, but we did find similar tables for all the variables. Our primary purpose is to determine the number of digits to which the data are given. We focus only on the continuous covariates, which DL label with "xx" in their dataset:

xxpolice police per 1000 residents – FBI's *Crime in the United States* [annual] – could not find numbers matching DL's, but one decimal is reported for "per 1000" categories. DL's dataset does not contain this variable but the log thereof, which is given to six decimals. This is unreasonable. A representative number of police per 1000 is 3.6. DL give the log of police in Alabama in 1997 as 1.179529. Now $\exp(1.179529) = 3.252842$ which rounds to 3.3. But $\exp(1.18) = 3.254374$ and, in fact, $\exp(1.2) = 3.320117$, both of which round to 3.3. So there is no harm in perturbing the log of police per 1000 residents in the fourth significant digit.

xxprison prisoners per 1000 residents – *Correctional Populations in the United States*, Bureau of Justice Statistics – gives number per 100,000, e.g., Alabama, 1998, 519 per 100,000 which translates to two decimals per 1000, e.g., 5.19. Curiously, DL Table III gives 2.83 as the overall average number of prisoners per 1000 residents, but has weighted this value by state population! Why a number per 1000 residents should be weighted for state population is unknown. Nonetheless, DL give six decimals for the log of this number, e.g., Alabama 1997 is 1.623693. Now, exp(1.623693) = 5.072, which rounds to 5.07. Since exp(1.623)= 5.068 and exp(1.624) = 5.073, both of which round to 5.072, there should be no harm in perturbing this variable in the fourth significant digit.

xxunemp state unemployment rate (percent) – *United States Statistical Abstract* – given to two digits, e.g., 4.1, while DL give this value equivalently to three decimals, e.g., Alabama 1997 = 0.051. So there should be no harm perturbing this variable in the fourth significant digit.

xxincome state personal income per capita ($1997) – *United States Statistical Abstract* (deflated by the 1997 CPI) – The nominal value is given in dollars not pennies, DL give the inflation-adjusted value to hundredths of a penny, e.g., log income for Alabama 1997 = 9.93727. Now exp(9.93727) = 20687.19. Since exp(9.93728) = 20687.4 and exp(9.93726)=20686.98, both of which round to 20687, there should be no harm perturbing in the seventh significant digit.

xxafdc15 AFDC generosity per recipient family – *United States Statistical Abstract* gives average monthly payment per family, which is multiplied by 12 and then deflated using the 1997 CPI. We could not find this variable, but let us assume that it is given in dollars not pennies; DL give to hundredths of a penny, e.g., Alabama 1997 = 2175.482. There should be no harm perturbing in the sixth significant digit.

xxbeer beer consumption per capita (gallons) – Beer Institute's *Brewer's Almanac* – given to one decimal, and DL do that. There should be no harm in perturbing in the third significant digit. – One-tenth or one-one-hundredth of a gallon should not matter.

xxpover poverty rate (percent below poverty level) – *United States Statistical Abstract* – given to one decimal, e.g., 5.3, and DL report this correctly. There should be no harm perturbing in the third significant digit.

xxefamurd this is the "effective abortion rate for murder" that DL computed. Its min and max are 4.56e-05 6.219058. We will perturb it in the fourth significant digit.

xxefaprop this is the "effective abortion rate for property crime" that DL computed. Its min and max are 0.0026488 9.765327. We will perturb it in the fourth significant digit.

xxefaviol this is the "effective abortion rate for violent crime" that DL computed. Its min and max are 0.000593 0.578164. We will perturb it in the fourth significant digit.

We do not perturb the dependent variable (but perhaps we should). When perturbing the data, we change the *i*-th digit by replacing it with a uniform draw from the integers 0–9.

We perform a BRB simulation where each continuous covariate is perturbed as specified above. While we should present BRB plots for each coefficient for each regression, to conserve space we present the three BRB plots (Fig. 1.3) for the parameter of interest in each regression, the coefficient on the "effective abortion rate".

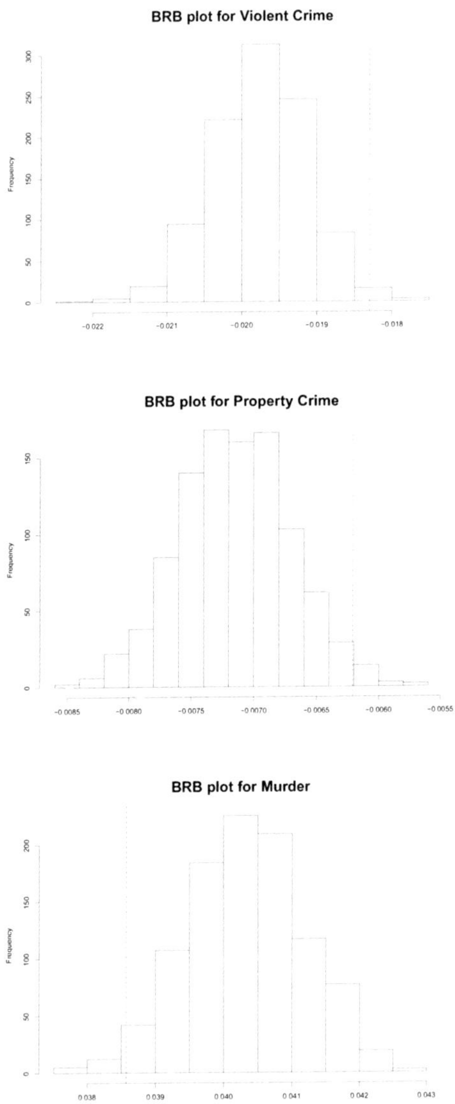

Fig. 1.3 BRB plots for "effective abortion rate" coefficients.

The conclusion is obvious and supports AW's contention that, as far as DL's regressions are concerned, the data were not up to the task of estimating the model.

1.8 Conclusions

We have shown that there is a definite need for econometric computing instruction in the economics profession: most users of econometric software, some developers of econometric software, and some authors of econometrics texts are unaware of the principles of accurate computing. We have shown that the typical econometric software package is not capable of being used for econometric computing: the programming languages contained in these packages are too simple to meet the needs of econometric computing. "R", by contrast, can be used for econometric computing; we give a standard computing exercise related to the Hilbert matrix as an example. We show how R can very easily be used to reproduce Beaton, Rubin and Barone's (1976) approach for investigating the numerical stability of a regression equation, and apply this method to the Donohue and Levitt (2000) abortion data. "R" is uniquely suited for teaching and applying the principles of statistical computing and econometric computing. Of special note in this regard is the fact that Monahan first supplied FORTRAN code to supplement his statistical computing text [24], and later added "R" code to do what the FORTRAN code already did.

Acknowledgements For useful comments, thanks to participants at Fordham University's 2009 "Conference on Quantitative Social Science Research Using R" and thanks also to Will Anderson, Charles Renfro and Houston Stokes.

References

1. Anderson, William and Martin T. Wells (2008), "Numerical Analysis in Least Squares Regression with an Application to the Abortion-Crime Debate,"*Journal of Empirical Legal Studies* **5**(4), 647-681
2. Beaton, A. E., D. B. Rubin and J. L. Barone (1976), "The acceptability of regression solutions: another look at computational accuracy", *JASA* **71**, 158-168
3. Belsley, David A. (1991), *Conditioning Diagnostics*, New York, Wiley
4. Brooks, Chris, Simon P. Burke and Gita Persand (2001), "Benchmarks and the Accuracy of GARCH Model Estimation," *International Journal of Forecasting* **17**(1), 45-56
5. Brooks, Chris, Simon P. Burke and Gita Persand (2003), "Multivariate GARCH Models: Software Choice and Estimation Issues," *Journal of Applied Econometrics* **18**(6), 725-734
6. Donohue, John J. and Steven D. Levitt (2001), "The Impact of Legalized Abortion on Crime," *Quarterly Journal of Economics* **116**(2), 379-420

7. Doornik, J. A. and R. J. O'Brien (2002), "Numerically Stable Cointegration Analysis," *Computational Statistics and Data Analysis* **41**, 185-193
8. Foote, Christopher L. and Christopher F. Goetz (2008), "The Impact of Legalized Abortion on Crime: Comment," *Quarterly Journal of Economics* **123**(1), 407-423
9. Fiorentini Gabrielle, Giorgio Calzolari and Lorenzo Panattoni (1996), "Analytic Derivatives and the Computation of GARCH Estimates," *Journal of Applied Econometrics* **11**(4), 399-417
10. Gentle, James (2007), *Matrix Algebra: Theory, Computation and Statistics*, New York, Springer
11. Gujarati, Damodar N. and Dawn C. Porter (2009), *Essentials of Econometrics, 4e*, New York, McGraw-Hill
12. Hammarling, Sven (2005), "An introduction to the quality of computed solutions" in B. Einarsson, editor, *Accuracy and Reliability in Scientific Computing*, 43-76, Philadelphia, SIAM
13. Longley, J. W. (1967), "An appraisal of least-squares programs from the point of view of the user," *Journal of the American Statistical Association* **62**, 819-841
14. McCullough, B. D. (1998), "Algorithm Choice for (Partial) Autocorrelation Functions," *Journal of Economic and Social Measurement* **24**(3/4), 265-278
15. McCullough, B. D. (1999), "Econometric Software Reliability: E-Views, LIMDEP, SHAZAM, and TSP," *Journal of Applied Econometrics* **14**(2), 191-202
16. McCullough, B. D. (2000), "Is it Safe to Assume that Software is Accurate?" *International Journal of Forecasting* **16**(3), 349-357
17. McCullough, B. D. (2004), "Wilkinson's Tests and Econometric Software," *Journal of Economic and Social Measurement* **29**(1-3), 261-270
18. McCullough, B. D. (2008), "Microsoft Excel's 'Not the Wichmann-Hill' Random Number Generators," *Computational Statistics and Data Analysis* **52**(10), 4587-4593
19. McCullough, B. D. (2009), "The Accuracy of Econometric Software" in *Handbook of Computational Econometrics*, 55-80, David A. Belsley and Erricos Kontoghiorghes, eds., New York, Wiley.
20. McCullough, B. D. (2009), "Testing Econometric Software" in *Handbook of Econometrics vol. 2*, 1293-1320, Terrence Mills and Kerry Patterson, eds., Palgrave-MacMillan, Basingbroke, UK
21. McCullough, B. D. and Charles G. Renfro (1999), "Benchmarks and Software Standards: A Case Study of GARCH Procedures," *Journal of Economic and Social Measurement* **25**(2), 59-71
22. McCullough, B. D. and H. D. Vinod (1999), "The Numerical Reliability of Econometric Software," *Journal of Economic Literature* **37**(2), 633-655
23. McCullough, B. D. and H. D. Vinod (2003), "Verifying the Solution from a Nonlinear Solver: A Case Study," *American Economic Review* **93**(3), 873-892
24. Monahan, John F. (2001), *Numerical Methods of Statistics*, Cambridge University Press, New York
25. Newbold, Paul, Christos Agiakloglou and John Miller (1994), "Adventures with ARIMA software," *International Journal of Forecasting* **10**(4), 573-581
26. Renfro, Charles G. (2009) *The Practice of Econometric Theory: An Examination of the Characteristics of Econometric Computation*, Springer, New York
27. Silk, Julian (1996), "Systems estimation: A comparison of SAS, SHAZAM and TSP," *Journal of Applied Econometrics* **11**(4), 437-450
28. Stokes, H. H. (2004), "On the Advantage of Using Two or More Econometric Software Systems to Solve the Same Problem," *Journal of Economic and Social Measurement* **29**, 307-320
29. Vinod, H. D. (2000), "Review of GAUSS for Windows, including its numerical accuracy," *Journal of Applied Econometrics* **15**, 211-222

30. Vinod, H. D. (2008), *Hands-On Intermediate Econometrics Using R: Templates for Extending Dozens of Examples*, World Scientific, Hackensack, NJ
31. Vinod, H. D. (2010), "Stress Testing of Econometric Results Using Archived Code," *Journal of Economic and Social Measurement* (to appear)
32. Yalta, Talha (2007), "Numerical Reliability of Econometric Software: The Case of GAUSS 8.0," *The American Statistician*, **61**, 262-268
33. Zeileis, Achim (2006), "Implementing a Class of Structural Change Tests: An Econometric Computing Approach," *Computational Statistics and Data Analysis* **50**, 2987-3008
34. Zellner, A. and H. Thornber (1966), "Computational Accuracy and Estimation of Simultaneous Equation Econometric Models," *Econometrica* **34**(3), 727-729

Chapter 2
Additive Models for Quantile Regression: An Analysis of Risk Factors for Malnutrition in India

Roger Koenker

Abstract This brief report describes some recent developments of the R `quantreg` package to incorporate methods for additive models. The methods are illustrated with an application to modeling childhood malnutrition in India.

Models with additive nonparametric effects offer a valuable dimension reduction device throughout applied statistics. In this paper we describe some recent developments of additive models for quantile regression. These methods employ the total variation smoothing penalties introduced in [9] for univariate components and [7] for bivariate components. We focus on selection of smoothing parameters including lasso-type selection of parametric components, and on post selection inference methods.

Additive models have received considerable attention since their introduction by Hastie and Tibshirani (1986, 1990). They provide a pragmatic approach to nonparametric regression modeling; by restricting nonparametric components to be composed of low-dimensional additive pieces we can circumvent some of the worst aspects of the notorious curse of dimensionality. It should be emphasized that we use the word "circumvent" advisedly, in full recognition that we have only swept difficulties under the rug by the assumption of additivity. When conditions for additivity are violated there will obviously be a price to pay.

Roger Koenker
Department of Economics, University of Illinois, Champaign, IL 61820, USA
e-mail: `rkoenker@uiuc.edu`

H.D. Vinod (ed.), *Advances in Social Science Research Using R*,
Lecture Notes in Statistics 196, DOI 10.1007/978-1-4419-1764-5_2,

2.1 Additive Models for Quantile Regression

Our approach to additive models for quantile regression and especially our implementation of methods in R is heavily influenced by Wood (2006, 2009) on the package mgcv. In some fundamental respects the approaches are quite distinct: Gaussian likelihood is replaced by (Laplacian) quantile fidelity, squared $\mathscr{L}_2$ norms as measures of the roughness of fitted functions are replaced by corresponding $\mathscr{L}_1$ norms measuring total variation, and truncated basis expansions are supplanted by sparse algebra as a computational expedient. But in many respects the structure of the models is quite similar. We will consider models for conditional quantiles of the general form

$$Q_{Y_i|x_i,z_i}(\tau|x_i,z_i) = x_i'\beta + \sum_{j=1}^{J} g_j(z_{ij}). \tag{1}$$

The nonparametric components g_j will be assumed to be continuous functions, either univariate, $\mathscr{R} \to \mathscr{R}$, or bivariate, $\mathscr{R}^2 \to \mathscr{R}$. We will denote the vector of these functions as $g = (g_1, \ldots, g_J)$. Our task is to estimate these functions together with the Euclidean parameter $\beta \in \mathscr{R}^K$, by solving

$$\min_{(\beta,g)} \sum \rho_\tau(y_i - x_i^T\beta + \sum g_j(z_{ij})) + \lambda_0 \parallel \beta \parallel_1 + \sum_{j=1}^{J} \lambda_j \bigvee(\nabla g_j) \tag{2}$$

where $\parallel \beta \parallel_1 = \sum_{k=1}^{K} |\beta_k|$ and $\bigvee(\nabla g_j)$ denotes the total variation of the derivative on gradient of the function g. Recall that for g with absolutely continuous derivative g' we can express the total variation of $g' : \mathscr{R} \to \mathscr{R}$ as

$$\bigvee(g'(z)) = \int |g''(z)| dz$$

while for $g : \mathscr{R}^2 \to \mathscr{R}$ with absolutely continuous gradient,

$$\bigvee(\nabla g) = \int \parallel \nabla^2 g(z) \parallel dz$$

where $\nabla^2 g(z)$ denotes the Hessian of g, and $\parallel \cdot \parallel$ will denote the usual Hilbert–Schmidt norm for this matrix. As it happens, solutions to (2) are piecewise linear with knots at the observed z_i in the univariate case, and piecewise linear on a triangulation of the observed z_i's in the bivariate case. This greatly simplifies the computations required to solve (2), which can now be written as a linear program with (typically) a very sparse constraint matrix consisting mostly of zeros. This sparsity greatly facilitates efficient solution of the resulting problem, as described in [8]. Such problems are efficiently solved by modern interior point methods like those implemented in the quantreg package.

2.2 A Model of Childhood Malnutrition in India

An application motivated by a recent paper [1] illustrates the full range of the models described above. As part of a larger investigation of malnutrition we are interested in determinants of children's heights in India. The data come from Demographic and Health Surveys (DHS) conducted regularly in more than 75 countries. We have 37,623 observations on children between the ages of 0 and 6. We will consider six covariates entering as additive nonparametric effects in addition to the response variable height: the child's age, gender, and months of breastfeeding, the mother's body mass index (bmi), age and years of education, and the father's years of education. Summary statistics for these variables appear in Table 2.1. There are also a large number of discrete covariates that enter the model as parametric effects; these variables are also summarized in Table 2.1. In the terminology of R categorical variables are entered as factors, so a variable like mother's religion that has five distinct levels accounts for four parameters.

Variable	Counts	Percent
csex		
male	19574	52.0
female	18049	48.0
ctwin		
singlebirth	37170	98.8
twin	453	1.2
cbirthorder		
1	11486	30.5
2	10702	28.4
3	6296	16.7
4	3760	10.0
5	5379	14.3
mreligion		
christian	3805	10.1
hindu	26003	69.1
muslim	6047	16.1
other	1071	2.8
sikh	697	1.9
mresidence		
urban	13965	37.1
rural	23658	62.9
deadchildren		
0	31236	83.0
1	4640	12.3
2	1196	3.2
3	551	1.5

Variable	Counts	Percent
wealth		
poorest	6625	17.6
poorer	6858	18.2
middle	7806	20.7
richer	8446	22.4
richest	7888	21.0
munemployed		
unemployed	24002	63.8
employed	13621	36.2
electricity		
no	10426	27.7
yes	27197	72.3
radio		
no	25333	67.3
yes	12290	32.7
television		
no	19414	51.6
yes	18209	48.4
refrigerator		
no	31070	82.6
yes	6553	17.4
bicycle		
no	19902	52.9
yes	17721	47.1
motorcycle		
no	30205	80.3
yes	7418	19.7
car		
no	36261	96.4
yes	1362	3.6

Table 2.1 Summary statistics for the response and continuous covariates

Ctab	Units	Min	Q1	Q2	Q3	Max
Chgt	cm	45.00	73.60	84.10	93.20	120.00
Cage	months	0.00	16.00	31.00	45.00	59.00
Bfed	months	0.00	9.00	15.00	24.00	59.00
Mbmi	kg/m^2	12.13	17.97	19.71	22.02	39.97
Mage	years	13.00	21.00	24.00	28.00	49.00
Medu	years	0.00	0.00	5.00	9.00	21.00
Fedu	years	0.00	2.00	8.00	10.00	22.00

Prior studies of malnutrition using data like the DHS have typically either focused on mean height or transformed the response to binary form and analyzed the probability that children fall below some conventional height cutoff. Nevertheless, it seems more natural to try to estimate models for some low conditional quantile of the height distribution. This is the approach adopted by FKH and the one we will employ here. It is also conventional in prior studies including FKH, to replace the child's height as response variable by a standardized Z-score. This variable is called "stunting" in the DHS data and it is basically just an age-adjusted version of height with age-specific location and scale adjustments. In our experience this preliminary adjustment is highly detrimental to the estimation of the effects of interest so we have reverted to using height itself as a response variable.

In R specification of the model to be estimated is given by the command

```
f <- rqss(Chgt ~ qss(Cage,lambda = 20) + qss(Mage, lambda = 80) +
    qss(Bfed,lambda = 80) + qss(Mbmi, lambda = 80) +
    qss(Medu, lambda = 80) + qss(Fedu, lambda = 80) +
    munemployed + csex + ctwin + cbirthorder + mreligion +
    mresidence + deadchildren + wealth + electricity + radio +
    television + refrigerator + bicycle + motorcycle + car,
    tau = .10, method = "lasso", lambda = 40, data = india)
```

The formula given as the first argument specifies each of the six non-parametric "smooth" terms. In the present instance each of these is univariate, and each requires specification of a λ determining its degree of smoothness. The remaining terms in the formula are specified as is conventional in other R linear model fitting functions like lm( ) and rq( ). The argument `tau` specifies the quantile of interest and `data` specifies the dataframe within which all of the formula variables are defined.

2.2.1 λ-Selection

A challenging task for any regularization problem like (2) is the choice of the λ parameters. Since we have seven of these the problem is especially daunting.

Following the suggestion originally appearing in Koenker, Ng and Portnoy [9] we relied on the Schwartz information criterion (SIC)-type criterion:

$$\mathrm{SIC}(\lambda) = n \log \hat{\sigma}(\lambda) + \tfrac{1}{2} p(\lambda) \log(n)$$

where $\hat{\sigma}(\lambda) = n^{-1} \sum_{i=1}^{n} \rho_\tau(y_i - \hat{g}(x,z))$, and $p(\lambda)$ is the effective dimension of the fitting model

$$\hat{g}(x,z) = x'\hat{\beta} + \sum_{j=1}^{J} \hat{g}_j(z).$$

The quantity $p(\lambda)$ is usually defined for linear least-squares estimators as the trace for pseudo projection matrix. The situation is somewhat similar for quantile regression fitting since we can simply compute the number of zero residuals for the fitted model. Recall that in unpenalized quantile regression fitting a p-parameter model yields precisely p zero residuals provided that the y_i's are in general position. This definition of $p(\lambda)$ can be viewed from a more unified perspective as consistent with the definition proposed by [11],

$$p(\lambda) = div(\hat{g}) = \sum_{i=1}^{n} \frac{\partial \hat{g}(x_i, z_i)}{\partial y_i},$$

see Koenker [5, p. 243]. A consequence of this approach to characterizing model dimension is that it is necessary to avoid "tied" responses; we ensure this by "dithering" the response variable. Heights measured to the nearest millimeter are replaced by randomly perturbed values by adding uniformly distributed "noise" $U[-0.05, 0.05]$.

Optimizing $SIC(\lambda)$ over $\lambda \in \mathbb{R}_+^7$ is still a difficult task made more challenging by the fact that the objective is discontinuous at points where new constraints become binding and previously free parameters vanish. The prudent strategy would seem to be to explore informally, trying to narrow the region of optimization and then resort to some form of global optimizer to narrow the selection. Initial exploration was conducted by considering all of the continuous covariate effects excluding the child's age as a group, and examining one-dimensional grids for λ's for this group, for the child's age, and the lasso λ individually. This procedure produced rough starting values for the following simulated annealing safari:

```
set.seed(1917)
malnu <- cbind(india, dChgt = dither(india$Chgt))

sic <- function(lam){
a <- AIC(rqss(dChgt~csex+qss(cage,lambda=lam[1])+
    qss(mbmi,lambda=lam[2])+ qss(Bfed,lambda=lam[3])+
    qss(Mage,lambda=lam[4])+ qss(Medu,lambda=lam[5])+
    qss(Fedu,lambda=lam[6])+ csex + ctwin+cbirthorder+
    munemployed+mreligion+mresidence + wealth+electricity+
    radio+television+refrigerator+bicycle+motorcycle+car,
    tau=0.1, method="lasso", lambda=lam[7], data=malnu), k=-1)
```

```
print(c(lam,a))
a
}
g <- optim(c(20,80,80,80,80,80,20),sic,method="SANN",
     control=list(maxit=1000,temp=5000, trace=10,REPORT=1))
```

Each function evaluation takes about 7 seconds, so 1000 steps of the simulated annealing algorithm required about 2 hours. The "solution" yielded

```
$par
[1] 16.34 67.92 78.49 85.05 77.81 82.51 17.63
$value
[1] 245034.0
```

Thus, the original starting values proved to be somewhat vindicated. We would not claim that the "solutions" produced by this procedure are anything but rough approximations. However, in our experience choosing λ's anywhere in a moderately large neighborhood of this solution obtained this way yields quite similar inferential results that we will now describe.

2.2.2 Confidence Bands and Post-Selection Inference

Confidence bands for nonparametric regression introduce some new challenges. As with any shrinkage type estimation method there are immediate questions of bias. How do we ensure that the bands are centered properly? Bayesian interpretation of the bands as pioneered by [15] and [12] provides some shelter from these doubts. For our additive quantile regression models we have adopted a variant of the Nychka approach as implemented by Wood [17] in the mgcv package.

As in any quantile regression inference problem we need to account for potential heterogeneity of the conditional density of the response. We do this by adopting Powell's [14] proposal to estimate local conditional densities with a simple Gaussian kernel method.

The pseudo design matrix incorporating both the lasso and total variation smoothing penalties can be written as

$$\tilde{X} = \begin{bmatrix} X & G_1 & \cdots & G_J \\ \lambda_0 H_K & 0 & \cdots & 0 \\ 0 & \lambda_1 P_1 & \cdots & 0 \\ \vdots & \cdots & \ddots & \vdots \\ 0 & 0 & \cdots & \lambda_j P_J \end{bmatrix}.$$

Here X denotes the matrix representing the parametric covariate effects, the G_j's represent the basis expansion of the g_j functions, $H_K = [0 : I_K]$ is the penalty contributions from the lasso excluding any penalty on the intercept and the P_j terms represent the contribution from the penalty terms on

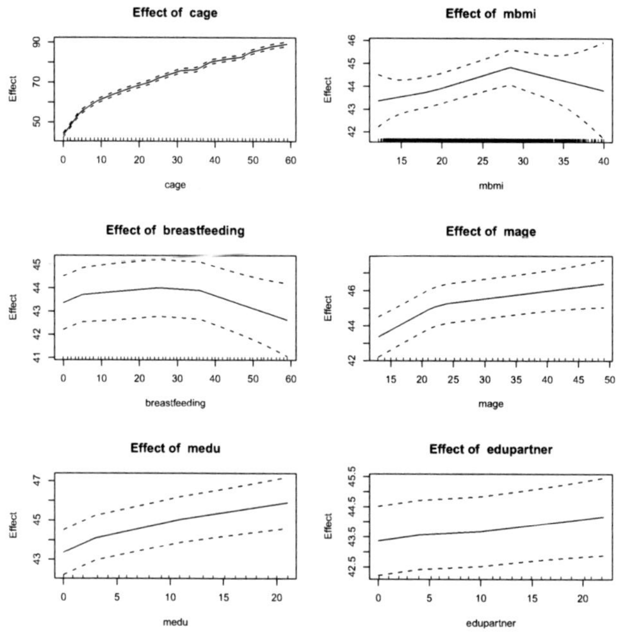

Fig. 2.1 Smooth covariate effects on children's heights with pointwise confidence bands.

each of the smoothed components. The covariance matrix for the full set of parameters, $\theta = (\beta^\top, \gamma_1^\top, \ldots, \gamma_J^\top)^\top$, is given by the sandwich formula,

$$V = \tau(1-\tau)(\tilde{X}^\top \Psi \tilde{X})^{-1}(\tilde{X}^\top \tilde{X})^{-1}(\tilde{X}^\top \Psi \tilde{X})^{-1}$$

where Ψ denotes a diagonal matrix with the first n elements given by the local density estimates,

$$\hat{f}_i = \phi(\hat{u}_i/h)/h,$$

$\hat{u}_i$ is the ith residual from the fitted model, and h is a bandwidth determined by one of the usual built-in rules. The remaining elements of the Ψ diagonal corresponding to the penalty terms are set to one.

Pointwise confidence bands can be easily constructed given this matrix V. A matrix D representing the prediction of g_j at some specified plotting points $z_{ij} : i = 1, \ldots, m$ is first made, then we extract the corresponding chunk of the matrix V, and compute the estimated covariance matrix of the vector $D\theta$. Finally, we extract the square root of the diagonal of this matrix. The only slight complication of this strategy is to remember that the intercept should be appended to each such prediction and properly accounted for in the extraction of the covariance matrix of the predictions.

To illustrate the use of these confidence bands, Fig. 2.1 shows the six estimated smoothed covariate effects and the associated confidence bands. This plot is produced by refitting the model with the selected λ's, calling the fitted model object `fit` and then using the command

```
plot(fit, bands = TRUE, page = 1)
```

Clearly the effect of age and the associated growth curve is quite precisely estimated, but the remaining effects show considerably more uncertainty. Mother's BMI has a positive effect up to about 30 and declines after that, similarly breastfeeding is advantageous up until about 30 months, and then declines somewhat. (Some breastfeeding after 36 months is apparently quite common in India as revealed by the DHS survey.)

What about inference on the parametric components of the model? We would certainly like to have some way to evaluate the "significance" of the remaining parametric coefficients in the model. Again, bias effects due to shrinkage create some serious doubts, and from a strict frequentist viewpoint these doubts may be difficult to push aside. See for example the recent work of [13]. However, a Bayesian viewpoint may again rescue the naive application of the covariance matrix estimate discussed above. When we employ this covariance matrix to evaluate the parametric component of the model, we obtain the following table from R using the usual `summary(fit)` command.

```
Parametric coefficients:
                     Estimate Std. Error  t value Pr(>|t|)
(Intercept)         4.336e+01  5.753e-01   75.382  < 2e-16 ***
csexfemale         -1.405e+00  4.516e-02  -31.110  < 2e-16 ***
ctwintwin          -6.550e-01  2.504e-02  -26.157  < 2e-16 ***
cbirthorder2       -6.492e-01  4.411e-02  -14.719  < 2e-16 ***
cbirthorder3       -9.491e-01  4.246e-02  -22.355  < 2e-16 ***
cbirthorder4       -1.437e+00  4.013e-02  -35.807  < 2e-16 ***
cbirthorder5       -2.140e+00  3.975e-02  -53.837  < 2e-16 ***
munemployedemployed 9.753e-02  4.453e-02    2.190 0.028532 *
mreligionhindu     -2.111e-01  4.185e-02   -5.043 4.61e-07 ***
mreligionmuslim    -1.957e-01  3.991e-02   -4.904 9.42e-07 ***
mreligionother     -3.934e-01  3.005e-02  -13.090  < 2e-16 ***
mreligionsikh      -2.353e-13  2.766e-02 -8.5e-12 1.000000
mresidencerural     1.465e-01  4.357e-02    3.363 0.000773 ***
wealthpoorer        2.126e-01  4.374e-02    4.861 1.17e-06 ***
wealthmiddle        5.880e-01  4.230e-02   13.899  < 2e-16 ***
wealthricher        8.368e-01  3.999e-02   20.924  < 2e-16 ***
wealthrichest       1.358e+00  3.540e-02   38.367  < 2e-16 ***
electricityyes      2.414e-01  4.345e-02    5.556 2.78e-08 ***
radioyes            4.073e-02  4.530e-02    0.899 0.368547
televisionyes       1.793e-01  4.378e-02    4.096 4.21e-05 ***
refrigeratoryes     1.289e-01  3.969e-02    3.247 0.001168 **
bicycleyes          3.940e-01  4.489e-02    8.778  < 2e-16 ***
motorcycleyes       1.764e-01  4.193e-02    4.207 2.60e-05 ***
caryes              3.633e-01  3.214e-02   11.303  < 2e-16 ***
```

There are a number of peculiar aspects to this table. Somewhat surprisingly, our "optimal" choice of the lasso λ of 17.63 only zeros out one coefficient

— the effect of the relatively small minority of sikhs. For all the remaining coefficients the effect of the lasso shrinkage is to push coefficients toward zero, *but also to reduce their standard errors.* The implicit prior represented by the lasso penalty acts as data augmentation that improves the apparent precision of the estimates. Whether this is regarded as a Good Thing is really questionable. To contrast the conclusions drawn from this table with somewhat more conventional methods, we have reestimated the model maintaining the smoothing λ's at their "optimized" values, but setting the lasso λ to zero.

```
Parametric coefficients:
                    Estimate Std. Error t value Pr(>|t|)
(Intercept)         43.51139    0.64391  67.574  < 2e-16 ***
csexfemale          -1.44232    0.08421 -17.128  < 2e-16 ***
ctwintwin           -0.86987    0.34680  -2.508  0.01214 *
cbirthorder2        -0.76125    0.10883  -6.995 2.70e-12 ***
cbirthorder3        -1.13288    0.14098  -8.036 8.88e-16 ***
cbirthorder4        -1.60645    0.18238  -8.808  < 2e-16 ***
cbirthorder5        -2.34391    0.20206 -11.600  < 2e-16 ***
munemployedemployed  0.09254    0.09348   0.990  0.32221
mreligionhindu      -0.42625    0.15390  -2.770  0.00561 **
mreligionmuslim     -0.50185    0.18902  -2.655  0.00793 **
mreligionother      -0.76162    0.25700  -2.963  0.00304 **
mreligionsikh       -0.39472    0.39786  -0.992  0.32114
mresidencerural      0.23299    0.10362   2.248  0.02456 *
wealthpoorer         0.45847    0.15372   2.982  0.00286 **
wealthmiddle         0.89591    0.17073   5.248 1.55e-07 ***
wealthricher         1.23945    0.20023   6.190 6.07e-10 ***
wealthrichest        1.83644    0.25340   7.247 4.33e-13 ***
electricityyes       0.14807    0.13215   1.120  0.26253
radioyes             0.01751    0.09701   0.180  0.85679
televisionyes        0.16862    0.12103   1.393  0.16359
refrigeratoryes      0.15100    0.14808   1.020  0.30787
bicycleyes           0.42391    0.08897   4.764 1.90e-06 ***
motorcycleyes        0.20167    0.13193   1.529  0.12637
caryes               0.49681    0.23161   2.145  0.03196 *
```

This table is obviously quite different: coefficients are somewhat larger in absolute value and more importantly standard errors are also somewhat larger. The net effect of removing the lasso "prior" is that many of the coefficients that looked "significant" in the previous version of the table are now of doubtful impact. Since we regard the lasso penalty more as an expedient model selection device rather than an accurate reflection of informed prior opinion, the latter table seems to offer a more prudent assessment of the effects of the parametric contribution to the model. A natural question would be: does the refitted model produce different plots of the smooth covariate effects? Fortunately, the answer is no, as replotting Fig. 2.1 with the unlasso'ed parametric fit yields a figure that is almost indistinguishable from the original.

Most of the estimated parametric effects are unsurprising: girls are shorter than boys even at the 10th percentile of heights, children later in the birth order tend to be shorter, mothers who are employed and wealthier have taller children, religious differences are very small, and some household capital stock variables have a weak positive effect on heights, even after the categorical wealth variable is accounted for.

The `summary(fit)` command also produces F-tests of the joint significance of the nonparametric components, but we will defer the details of these calculations. A further issue regarding these nonparametric components would be the transition from the pointwise confidence bands that we have described above to uniform bands. This topic has received quite a lot of attention in recent years, although the early work of [4] has been crucial. Recent work by [10] has shown how to adapt the Hotelling approach for the same GAM models in the Wood `mgcv` package. It appears that similar methods can be adapted to `rqss` fitting; I hope to report on this in future work.

Acknowledgements This research was partially supported by NSF grant SES-08-50060. I would like to express my appreciation to Ying Li for excellent research assistance. All of the methods described here have been implemented in version 4.42 of the `quantreg` package for `R`, [6].

References

1. Fenske, N., Kneib, T., Hothorn, T.: Identifying risk factors for severe childhood malnutrition by boosting additive quantile regression (2008). Preprint
2. Hastie, T., Tibshirani, R.: Generalized additive models. Statistical Science **1**, 297–310 (1986)
3. Hastie, T., Tibshirani, R.: Generalized Additive Models. Chapman-Hall (1990)
4. Hotelling, H.: Tubes and spheres in n-space and a class of statistical problems. American J of Mathematics **61**, 440–460 (1939)
5. Koenker, R.: Quantile Regression. Cambridge U. Press, London (2005)
6. Koenker, R.: quantreg: A quantile regression package for r (2009). `http://cran.r-project.org/src/contrib/PACKAGES.html#quantreg`
7. Koenker, R., Mizera, I.: Penalized triograms: total variation regularization for bivariate smoothing. J. Royal Stat. Soc. (B) **66**, 145–163 (2004)
8. Koenker, R., Ng, P.: A frisch-newton algorithm for sparse quantile regression. Mathematicae Applicatae Sinica **21**, 225–236 (2005)
9. Koenker, R., Ng, P., Portnoy, S.: Quantile smoothing splines. Biometrika **81**, 673–680 (1994)
10. Krivobokova, T., Kneib, T., Claeskens, G.: Simultaneous confidence bands for penalized spline estimators (2009). Preprint
11. Meyer, M., Woodroofe, M.: On the degrees of freedom in shape-restricted regression. Annals of Stat. **28**, 1083–1104 (2000)
12. Nychka, D.: Bayesian confidence intervals for smoothing splines. J. of Am. Stat. Assoc. **83**, 1134–43 (1983)
13. Pötscher, B., Leeb, H.: On the distribution of penalized maximum likelihood estimators: The lasso, scad and thresholding. J. Multivariate Analysis (2009). Forthcoming

14. Powell, J.L.: Estimation of monotonic regression models under quantile restrictions. In: W. Barnett, J. Powell, G. Tauchen (eds.) Nonparametric and Semiparametric Methods in Econometrics. Cambridge U. Press: Cambridge (1991)
15. Wahba, G.: Bayesian "confidence intervals" for the cross-validated smoothing spline. J. Royal Stat. Soc. (B) **45**, 133–50 (1983)
16. Wood, S.: Generalized Additive Models: An Introduction with R. Chapman-Hall (2006)
17. Wood, S.: mgcv: Gams with gcv/aic/reml smoothness estimation and gamms by pql (2009). `http://cran.r-project.org/src/contrib/PACKAGES.html#mgcv`

Chapter 3
Toward Better R Defaults for Graphics: Example of Voter Turnouts in U.S. Elections

Andrew Gelman

R is great, and it is a pleasure to contribute to a volume on its practical applications. As part of my goal to make R even better, I want to discuss some of its flaws.

I will skip past computational issues (the difficulty of working with S4 objects, the awkwardness of the links to R and C, the accretion of ugly exception-handling code in many R functions, and well-known problems with speed and memory usage), and move to some issues I have had with graphics in R.

My #1 problem with R graphics is that its defaults do not work well for me. I always end up wrapping my code with extra instructions to have it do what I want. The simple graph (Fig. 3.1) represents a recent example. And

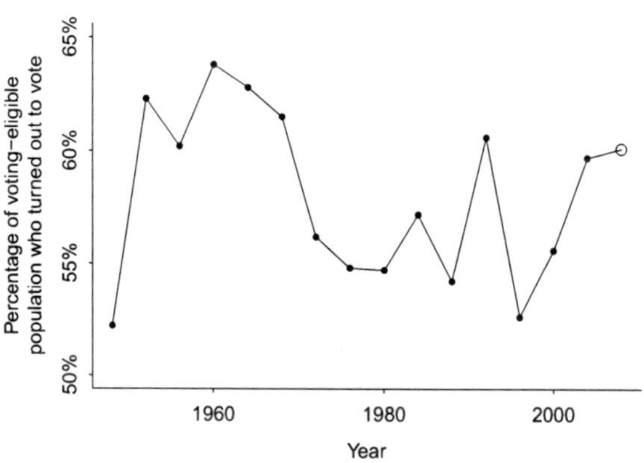

Fig. 3.1 Voter turnout in presidential elections, 1948-2008.

Andrew Gelman
Department of Statistics and Department of Political Science,
Columbia University, New York, NY 10027, USA e-mail: `gelman@stat.columbia.edu`

H.D. Vinod (ed.), *Advances in Social Science Research Using R*, 35
Lecture Notes in Statistics 196, DOI 10.1007/978-1-4419-1764-5_3,

here is the paragraph of ugly code needed to create the graph:

```
turnout.year <- seq (1948,2008,4)
turnout.VEP <- c(.522,.623,.602,.638,.628,.615,.562,.548,
  .542,.552,.528,.581,.517,.542,.601,.617)
pdf ("c:/temp/turnout.pdf", height=4, width=5.5)
n <- length(turnout.year)
par (mar=c(3,4,2,0), tck=-.01, mgp=c(2,.5,0))
plot (turnout.year, turnout.VEP, type="l", xlab="Year",
  ylab="Percentage of voting-eligible\npopulation who
  turned out to vote", xaxt="n", yaxt="n", bty="l",
  ylim=c(.50,.65))
points (turnout.year[1:(n-1)], turnout.VEP[1:(n-1)],
  pch=20)
points (turnout.year[n], turnout.VEP[n], pch=21, cex=1.2)
axis (1, seq (1960,2000,20))
yticks <- seq (.50,.65,.05)
axis (2, yticks, paste (yticks*100,"%",sep=""))
mtext ("Voter turnout in presidential elections, 1948-2008",
  line=1)
dev.off()
```

The script to create this graph should be one line, right? What is all this other stuff for? Two lines to set up the png close it when it is done; that is fair. A line at the end to give the graph a title. The plotting call itself takes up a couple of lines to specify the axis labels. An extra line of code to put a circle to emphasize the last data point. OK, so far, so good.

But what about the rest—unreadable machine-language-like code with tags such as "mar", "mgp", "cex", "sep", and the rest? This extra code is mostly there to fix ugly defaults of R graphics, including the following:

- Tick marks that are too big. (They're OK on the windows graphics device, but when I make my graphs using postscript() or png() or pdf()—and it is good practice to do this—then it is necessary to set them much smaller so that they are not so big.)
- Axis numbers that are too closely spaced together.
- Axis labels too far from the axes.
- Character sizes that are way too small or way too big for the size of the graph. (This really starts to become a problem if you want to resize a plot.)

Just try making the above graph using the default settings and you will see what I mean. I routinely set xaxt="n", yaxt="n", and type="n" so that I can set everything manually. Some of these choices would be difficult to automate (for example, "45%" rather than the ugly "0.45" on the y-axis), but others (such as too-frequent tick marks, too-wide margins, and character sizes that do not work with the graph size) just baffle me.

Yes, I know I can write my own plotting functions and set my own defaults, but how many people actually do this?

I will also take this opportunity to mention a few other defaults which annoy me but do not happen to show up in the above example:

- Histogram bins that are too wide. (Whenever I make a histogram, I set the bin locations manually to get more resolution.)
- Lines that are too fat. (I always use lwd=.5 and I wish there was an even narrower setting.)
- Margins too big between plots. (In the default setting, you get tiny squares surrounded by oceans of white space. I always have to set the margins to be much smaller.)
- For all-positive variables, when the axis includes 0, it typically goes negative (unless you remember to set yaxs="i"). And for variables such as percentages that go from 0 to 1, the axis can go below 0 and above 1 (again, unless you alter the default settings). I know this choice can be justified sometimes (see page 32 of *The Elements of Graphing Data*, by William S. Cleveland [1, p. 32]), but in my experience these extended axes are more of a hindrance than a help for variables with natural constraints.

I have heard there are better tools out there for R graphics (for example, see Frank Harrell's hmisc package [2], Hadley Wickham's gg2plot [5], and the recent book *R Graphics* by Paul Murrell [3]), and I would not be surprised if the problems mentioned above get fixed soon. Nonetheless, the issues are here right now, and I think they illustrate some deeper statistical ideas that are not often studied.

Let me briefly discuss this last issue—axes that go beyond their logical range—because it illustrates a connection to deeper statistical ideas. Again, the R fix is easy enough (just set ylim=c(0,1),yaxs="i"), but the more interesting question is how to get better defaults. All these little fixes add up in time and effort, and most users will not go beyond the default anyway, which is why we see these little mistakes all over the place.

Axes that extend beyond the possible range of the data are not simply an issue of software defaults but reflect something more important, and interesting, which is the way in which graphics objects are stored on the computer.

R (and its predecessor, S) is designed to be an environment for data analysis, and its graphics functions are focused on plotting data points. If you are just plotting a bunch of points, with no other information, then it makes sense to extend the axes beyond the extremes of the data, so that all the points are visible. But then, if you want, you can specify limits to the graphing range (for example, in R, xlim=c(0,1),ylim=c(0,1)). The defaults for these limits are the range of the data.

What R does not allow, though, are logical limits: the idea that the space of the underlying distribution is constrained. Some variables have no constraints, others are restricted to be nonnegative, others fall between 0 and 1, others are integers, and so forth. R (and, as far as I know, other graphics packages) just treats data as lists of numbers. You also see this problem with discrete variables; for example, when R is making a histogram of a variable that takes on the values 1, 2, 3, 4, 5, it does not know to set up the bins at the correct places, instead setting up bins from 0 to 1, 1 to 2, 2 to 3, etc., making it nearly impossible to read sometimes.

What I think would be better is for every data object to have a "type" attached: the type could be integer, nonnegative integer, positive integer, continuous, nonnegative continuous, binary, discrete with bounded range, discrete with specified labels, unordered discrete, continuous between 0 and 1, etc. If the type is not specified (i.e., NULL), it could default to unconstrained continuous (thus reproducing what is in R already). Graphics functions could then be free to use the type; for example, if a variable is constrained, one of the plotting options (perhaps the default, perhaps not) would be to have the constraints specify the plotting range.

Lots of other benefits would flow from this, I think, and that is why we are doing this in our 'mi' package for multiple imputation (in collaboration with Jennifer Hill, Yu-Sung Su, and others [4]). But the basic idea is not limited to any particular application; it is a larger point that data are not just a bunch of numbers; they come with structure.

I know that the R development community recognizes both of my points—the problems with many of the graphical defaults and the usefulness of variable types. Much of the challenge is in the implementation. Nonetheless, it may be helpful to lay out the connections between these practical and theoretical issues, which may help point a way toward more effective default settings in statistical graphics.

Acknowledgements The data on turnout as a proportion of the voting-eligible population were compiled by political scientist Michael McDonald at George Mason University and appear at http://elections.gmu.edu/voter_turnout.htm. I thank Hrishikesh Vinod for inviting this article and Hadley Wickham and Karl Ove Hufthammer for helpful comments.

References

1. Cleveland, W.S.: The Elements of Graphing Data. Hobart Press (1994)
2. Harrell, F.E., with contributions from many other users: Hmisc: Harrell Miscellaneous (2007)
3. Murrell, P.: R Graphics (Computer Science and Data Analysis). Chapman & Hall/CRC (2005)
4. Su, Y., Gelman, A., Hill, J., Yajima, M.: Multiple Imputation with Diagnostics (mi) in R: Opening Windows into the Black Box. Journal of Statistical Software. Forthcoming
5. Wickham, H.: ggplot2 : elegant graphics for data analysis. Springer (2009)

Chapter 4
Superior Estimation and Inference Avoiding Heteroscedasticity and Flawed Pivots: R-example of Inflation Unemployment Trade-Off

H. D. Vinod

Abstract We use a new solution to the heteroscedastic regression problem while avoiding the so-called incidental parameters (inconsistency) problem by using recently discovered maps from the time domain to numerical values domain and back. This involves a parsimonious fit for sorted logs of squared fitted residuals. Dufour [9] showed that inference based on Fisher's pivot (dividing by standard errors) can be fundamentally flawed for deep parameters of genuine interest to policymakers. Hence, we use Godambe's [12] pivot, which is always a sum of T items and asymptotically subject to the central limit theory. We provide R functions to implement the ideas using the Phillips curve trade-off between inflation and unemployment for illustration. The Appendix discusses numerical methods to correct for general ARMA errors with an illustration of ARMA(4,3).

Key words: feasible generalized least squares; specification robust; smoothness; simulation; consistency

H. D. Vinod
Fordham University, Bronx, NY 10458, USA e-mail: `Vinod@fordham.edu`

Prepared for presentation at: *Conference on Quantitative Social Science Research Using R*, Fordham University, New York, June 18-19, 2009.

H.D. Vinod (ed.), *Advances in Social Science Research Using R*, Lecture Notes in Statistics 196, DOI 10.1007/978-1-4419-1764-5_4,

4.1 Introduction

Vinod's [29, Sect. 9.3.1] recent textbook mentions two-way maps between the values domain and time domain (denoted here by t-dom ↔ v-dom) based on [27, 28] and [30] developed in the context of maximum entropy bootstrap (meboot) package in R software. This paper describes an application of the t-dom ↔ v-dom maps to construct a parsimonious estimate of the heteroscedastic variances of unknown form in the context of the usual model for the generalized least squares (GLS) from textbooks [7, Chap. 7] and [29, Chap. 10]:

$$y = X\beta + \varepsilon,\ \ E\varepsilon|X = 0,\ \ E\varepsilon\varepsilon'|X = \sigma^2\Omega, \tag{4.1}$$

where X is $T \times p$, β is $p \times 1$, y and ε are $T \times 1$. The $X'X$ matrix and the large $T \times T$ matrix Ω are both assumed to be positive definite and known with $T > p$.

The feasible GLS (FGLS) problem under unknown form heteroscedasticity was rigorously studied long ago by Eicker [10]. Besides $E\varepsilon_t = 0$, he assumes that the individual errors satisfy $0 < E\varepsilon_t^2 = \Omega_{tt} < \infty$, with distribution functions (dfs) which are neither assumed to be known, nor identical for all $t \in 1,2,\ldots,T$. They are assumed, however, to be elements of a certain set F of dfs. Eicker denotes by $\mathscr{F}(F)$ the set of all sequences occurring in the regressions and a parameter point as a sequence of $\mathscr{F}(F)$.

Result 4.1. If ε_t are independent and identically distributed (iid), it is well known that the law of large numbers implies that the empirical distribution function based on the order statistics $\varepsilon_{(t)}$ gives a consistent estimate of the distribution function of errors.

Result 4.2 (Weierstrass, 1885). Since unknown error variances Ω_{tt} are defined at each $t \in 1,2,\ldots,T$, we can write them as a real-valued function $f(t)$ of t defined over a finite interval. Then the Weierstrass approximation theorem states that for every $\varepsilon > 0$, there exists a polynomial function $g(t)$ over $\Re$ such that the supremum norm satisfies: $\sup||\mathrm{f} - \mathrm{g}|| < \varepsilon$.

Since (4.1) relaxes the identical distribution part of the iid assumption, it cannot achieve the consistency in Result 4.1 without further assumptions. While Weierstrass Result 4.2 shows that a polynomial approximation exists, this paper uses the t-dom ↔ v-dom maps to suggest a new, simple and practical method of finding $g(t)$ using a new set of assumptions A1 to A3 on the set $\mathscr{F}(F)$. A general function $\dot{g}(X,t)$ to approximate $f(t)$ by using nonparametric estimates, such as those in [24], can be used, but left for future work.

The OLS estimator is $b = (X'X)^{-1}X'y$. It has Eicker's covariance matrix $\mathrm{Cov}(b) = \sigma^2(X'X)^{-1}[X'\Omega X](X'X)^{-1}$, where σ^2 is commonly estimated by $s^2 = (y - Xb)'(y - Xb)/(T - p)$. It is convenient to simplify the expressions in the sequel by merging the scalar σ^2 inside our Ω.

Ignoring the Ω matrix makes OLS estimator b inefficient and standard errors $SE(b)$ potentially misleading. However, direct estimation of T diagonal elements of Ω from all T squared residuals s_{tt}, using only T data points, faces the well-known "incidental parameters" (inconsistency) problem. The basic solution involves using $\hat{\Omega} = \Omega(\hat{\phi})$, where the number of parameters in the ϕ vector are far less than T. Carroll [2] assumes that the variance function Ω_{tt} is a smooth function, in the sense that it has a continuous first derivative, in a time domain neighborhood, which can also be justified by Result 4.2.

Assuming a consistent estimate $\hat{\Omega}$ is available, the feasible GLS (FGLS) estimator of β is given by

$$b_{FGLS} = [X'\hat{\Omega}^{-1}X]^{-1}X'\hat{\Omega}^{-1}y, \text{ with } \mathrm{Cov}(b_{FGLS}) = [X'\hat{\Omega}^{-1}X]^{-1}. \tag{4.2}$$

This paper is concerned with correcting for heteroscedasticity of the unknown form considered by many authors, such as those surveyed in [18]. Some have attempted to solve the heteroscedasticity problem by going outside the FGLS class by modifying the minimand. Professor C. R. Rao suggested a variance components model on assuming fewer components than T and a quadratic form $y'Ay$ to estimate them. His minimand is the Euclidean norm of the difference between the true quadratic form and its estimator leading to minimum norm quadratic estimation (MINQUE). The large literature inspired by MINQUE is reviewed in [23].

The literature dealing with rank-based R-estimates of the regression model minimizing (Jaeckel's) dispersion function involving weighted ranks of errors is also large. Dixon and McKean [8] analyze the heteroscedastic linear model using the dispersion of residuals defined by $\sum_{t=1}^{T} w(t)R(\hat{\varepsilon}_t)(\hat{\varepsilon}_t)$, where $w(t)$ are weights and $R(.)$ represents ranks. The design of Dixon and McKean's simulation using monkey data and 10% contaminated Normal density shows a focus on solving a heteroscedasticity problem primarily caused by outliers and influential observations. Our proposal retains the usual score function leading to FGLS estimation: $X'\Omega^{-1}(y - X\beta)$.

Auxiliary variables needed for efficient estimation under heteroscedasticity of the unknown form in the current literature are mostly constructed from the matrix of regressors X wedded to the time domain. For example, White [31] and Cragg [5] use squares and cross-products of columns of X, whereas when Cragg constructs $\delta_t = \sigma_t^2 - \bar{\sigma}^2$ based on the deviation of residual variances from their average, he is implicitly recognizing that what matters for heteroscedasticity is their numerical magnitude.

We define the time domain (t-dom) as the place where our observable real numbers in $y, X, \varepsilon, \hat{\varepsilon}$ (e.g., time series) and functions using them are located. The values domain (v-dom) contains order statistics of the same real numbers (ordered from the smallest to the largest) and functions using them. After showing in the next section that magnitudes are best studied in the values domain, this paper suggests a new nonstochastic auxiliary variable constructed from a simple sequence of integers (1,2,...). Our method for correcting for heteroscedasticity is much easier to implement than what is currently available.

The outline of the remaining paper is as follows. Section 4.2 discusses the new heteroscedasticity cum heterogeneity efficient (HE) estimation. Section 4.3 reports a simulation. Section 4.4 contains an example. Section 4.6 has a summary and our final remarks. If autocorrelation is suspected, Vinod [29, Chap. 10] argues that one should make the autocorrelation correction before any heteroscedasticity correction, using parsimonious expressions for the $\hat{\Omega}$ matrix when the errors in (4.1) follow an ARMA process. The Appendix illustrates new corrections for ARMA(4,3) errors by numerical methods.

We are assuming that on performing tests for heteroscedasticity the researcher has already decided to correct for it by using FGLS estimation. All methods discussed in this paper are readily implemented by using the R software, similar to [29].

4.2 Heteroscedasticity Efficient (HE) Estimation

Let $V = \Omega^{-1/2}$ and $\hat{V} = \hat{\Omega}^{-1/2}$ denote the square root matrix of the inverse of Ω and its estimated $\hat{\Omega}$ version, respectively. Now rewrite (4.1) and verify that

$$Vy = VX\beta + V\varepsilon,\ \ EV\varepsilon|X = 0,\ \ E[V\varepsilon\varepsilon'V'|X] = \sigma^2 I_T. \qquad (4.3)$$

Let $H = X(X'X)^{-1}X'$ be the usual hat matrix and let its diagonals be denoted by H_{tt}. It is well known that $\text{var}(\hat{u}_t) = E\hat{u}_t^2 = \Omega_{tt}(1 - H_{tt})$. Efficient estimation under heteroscedasticity tries to give a larger weight to observations having a lower $\text{var}(\hat{u}_t)$ than to those having a higher variance. Accordingly, the following estimates of Ω_{tt} are found in the literature:

(HC0): $s_{tt,0} = \hat{u}_t^2 = [y_t - E(y_t|X)]^2$,
(HC1): $s_{tt,1} = T\hat{u}_t^2/(T-p)$,
(HC2): $s_{tt,2} = \hat{u}_t^2/(1 - H_{tt})$,
(HC3): $s_{tt,3} = \hat{u}_t^2/(1 - H_{tt})^2$ and
(HC4): $s_{tt,4} = \hat{u}_t^2/(1 - H_{tt})^{\delta tt}$, where $\delta tt = \min(4, H_{tt}/\text{mean}(H_{tt}))$.

Note that HC0 is a proxy for the conditional scale of y_t, which is a nonnormal nonnegative random variable. Davidson and MacKinnon [7, p. 200] define HC1 to HC3 and suggest a slight preference for HC3 based on the jackknife. Cribari-Neto [6] suggests HC4. Our discussion includes computation of HC0 to HC4 under the generic name s_{tt}. Following Cook and Weisberg [4] we can use scaled residuals, $e = \hat{u}/s$, in place of $\hat{u}$ although this change does not seem to make a practical difference.

Since nonconstant diagonal *values* of Ω become obvious on reordering them, heteroscedasticity is easier to study and more meaningful in the values (numerical magnitudes) domain than in the time domain. For example, if $\max(\hat{u}^2)$ is a lot larger (>>) than $\min(\hat{u}^2)$, it stands out better in the v-domain. One should use formal tests to make sure we have heteroscedasticity before proceeding to correct for it. Our example uses Cook and Weisberg's [4] score test.

Before we proceed, note that we cannot rule out a locally perfect fit. Hence some T' (say) of the $\hat{u}_t^2$ values may well be zero. This creates a practical problem that their $\log(s_{tt})$ becomes $-\infty$. McCullough and Vinod [17] argue against the temptation to replace the T' zeros with suitably small numbers close to zero. Hence, we work with the remaining "good" observations, $T_g = T - T'$, while excluding the troublesome T' components from the following heteroscedasticity correction algorithm.

4.2.0.1 Map t-domain to v-domain

We construct a $T \times 2$ matrix W, having the first column containing the vector $\tau = (1,2,...,T)'$, and the second column containing s_{tt} (generic for $\hat{u}_t^2$, or any one of HC1 to HC4). Next, we sort the W matrix on the second column, ordering its elements from the smallest to the largest, while carrying the first column along during the sorting process. This finds the monotonic order statistics $s_{(tt)}$ belonging to the v-domain. In the second column, the T' elements (associated with perfect fit or zero residuals) occupying initial positions will be zero due to the sorting process. Now, the remaining T_g (good) elements will be nonzero, with well-defined logarithms.

Use a subscript "s" to denote the sorted version of the W matrix: $W_s = \{W_{s,i,j}\}$, where its elements for row i and column j bear the subscript (s,i,j). Denote its columns 1 and 2 as $\tau_s = W_{s,.,1}$ and $s_{(tt)} = W_{s,.,2}$, respectively, replacing the "i" by a dot to represent the entire range of "i" values. The sorted version τ_s will be useful later for the reverse map of Sect. 4.2.0.2.

Assumption 4.1 (smoothness of heteroscedasticity). Recall Eicker's $\mathscr{F}(F)$ set of all sequences for distribution functions. We assume that the ordered values behind the distribution functions satisfy $\Omega_{(tt)} = f(t)$, where $f(t)$ is a piecewise smooth function in the (numerical values) v-domain. The assumption implies that $f(t)$ has a positive and piecewise continuous derivative.

The Weierstrass approximation of Result 4.2 allows us to approximate Ω_{tt} by a polynomial in the time domain. Our t-dom $\leftrightarrow$ v-dom maps are (one-one onto) bijections, and will be shown to be linear transforms, since they boil down to premultiplication by a matrix similar to the matrix (4.9) or its inverse. Hence the existence of a polynomial approximation holds in both domains and the approximating polynomials can be mapped between the two domains, as we wish. Our Assumption 4.1 imposes a smoothness requirement on the ordered true values $\Omega_{(tt)}$ in the v-domain for a convenient polynomial approximation. We do not claim that sorting is *necessary* in each and every example. However, our polynomial is in powers of $\{t\}$ a monotonic subset of integers $(1, 2, \ldots)$. Hence monotonic values from the v-domain as the dependent variable are expected to permit a lower order h of a good Weierstrass approximating polynomial.

In financial economics, Ω_{tt} values often represent price volatility values which can bunch together in sets of low, medium or high volatility regimes, representing distinct segments of $f(t)$. Assumption 4.1 mentions "piecewise" smooth functions so that distinct smooth functions for low, medium and high volatility regimes are available, if needed.

Since variances $f(t) = \Omega_{(tt)} > 0$ must be positive, we approximate $f(t)$ by using an exponential link function in the terminology of general linear models. We define the approximating polynomial in the population as

$$g(t) = \exp\left[\sum_{k=0}^{h} \phi_k t^k\right] \quad \text{for } t = (T'+1), (T'+2), \ldots, T. \tag{4.4}$$

Defining $\{t\}$ from the above range, a sample estimate of $g(t) > 0$ is obtained from the following population regression involving only observable variables on the two sides:

$$\log\, g(t) = \log\, s_{(tt)} = \sum_{k=0}^{h} \phi_k t^k + \varepsilon_{s.t}, \tag{4.5}$$

where the integers in $\{t\}$ are obviously nonstochastic and exogenous to the model. We expect to choose the order h of the polynomial in $\{t\}$ after some trial and error, perhaps starting with a quintic ($h = 5$) and using the multiple R^2 of (4.5) adjusted for degrees of freedom as a guide. We also recommend graphs to assess the suitability of fitted shapes and the need for segments, if any, of our piecewise continuous function. Formally, the impact of sampling variation in the estimation of $g(t)$ by $\exp \hat{s}_{(tt)}$ is made asymptotically negligible by letting h increase with T at a suitable rate, while invoking Result 4.2. Let (4.5) represent a model after the suitable h is found.

Let X^τ denote the matrix of regressors in (4.5). The $(h+1)$ normal equations to obtain the OLS estimates of ϕ must be solved simultaneously and represent the following "moment condition" (in the GMM literature terminology) imposed in the values domain, which yields improved efficiency, Newey [18]:

$$E\left[X^{\tau\prime}\left(\log s_{(tt)} - \sum_{j=0}^{h} \phi_j t^j\right)\right] = 0. \tag{4.6}$$

This equation uses "normal equations" for (4.5) to write a corresponding quasi score function as our estimating function. It may be possible to show efficiency improvement with reference to the estimating function literature, without mentioning the "moment conditions" terminology from econometrics.

Assumption 4.2. The polynomial regression in (4.5) in the values domain satisfies the usual assumptions: validity of (4.5), $E(\varepsilon_{s,t}) = 0$, and satisfaction of Grenander conditions [14, p. 354] for convergence, regressors are uncorrelated with errors (exogenous), and $E\varepsilon_{s,t}\varepsilon'_{s,t} = \sigma^2_{s,t}I$.

Since all variables in (4.5) are monotonic, we expect a good fit. Other parts of Assumption 4.2 ensure that the following well-known lemma holds. The proof is omitted, since our exogenous regressors (based on a sequence of integers) readily satisfy full column rank and Grenander's conditions:

Lemma 4.1. *Let $\hat{\phi}$ denote the $(h+1) \times 1$ vector of ordinary least squares estimates of coefficients ϕ in (4.5). Let $\Rightarrow$ denote convergence in probability as $T \to \infty$. Assuming Assumption 4.2, we have $\hat{\phi} \Rightarrow \phi$ in the values domain, provided the form of the parent density is known.*

The unusual requirement that "the form of the parent density is known" is explained in Kendall and Stuart [15, Sect. 19.21] and arises here because we are working with ordered observations. Using the lemma we estimate the smooth function $f(t)$ by $\exp(\hat{s}_{(tt)})$, where $\hat{s}_{(tt)}$ is

$$\hat{s}_{(tt)} = \sum_{k=0}^{h} \hat{\phi}_k t^k. \tag{4.7}$$

Let us assume that the mean and variance of the parent density of errors exist. Now use the usual quasi-Normality arguments to state that we can assume that the density can be approximated by the Normal density. Then, by properties of OLS, $\hat{s}_{(tt)}$ obtained by a form of trend-fitting provides an unbiased (consistent) estimator of $g(t)$, with $\hat{s}_{(tt)} = s_{(tt)} + \hat{\varepsilon}_{s,t}$. Even though we are using up a count of only $h+1 << T$ parameters in estimating the parsimonious regression (4.5), we cannot guarantee that our estimate of $f(t)$ is consistent for an unknown parent density. Hence we use a device of deleting additional T'' observations, familiar from the spectral window methods, [20, p. 437] further explained in Sect. 4.2.0.2 below. These deletions will be a part of our Assumption 4.3 below.

The residuals $\hat{\varepsilon}_{s,t}$ of regression (4.7) are both positive and negative, and can be regarded as either inside or outside a "window" based on a tolerance constant. Accordingly, all $(s_{(tt)})$ values associated with $|\hat{\varepsilon}_{s,t}| > M$ for some tolerance constant $M > 0$ are regarded as "outliers," from the true smooth function of Assumption 4.1. Choosing appropriately small M we have some $T'' > 0$ outliers, permitting the deletion of additional T'' values of $s_{(tt)}$. The next paragraph and the following subsection show that a practitioner is not burdened with making the "appropriately small" choice of M.

The beauty of the v-domain is that we can omit $T' + T''$ elements, truncating the left-hand side of (4.5) with impunity. We can simply use the entire sequence $t = \{1, 2, ..., T\}$, instead of a shorter sequence, $t = \{T'+1, T'+2, ..., T-T''\}$ created by the deletions on the right-hand side of (4.7) to get the right number of estimates in the time domain. Of course, it needs the reverse map from values to time domain, described next.

4.2.0.2 The Reverse Map: t-dom ← v-dom

The true unknown smooth function $f(t)$ is approximated by $\exp(\hat{s}_{(tt)})$ in the v-domain. We still need to map this into the time domain to get the diagonals Ω_{tt} of Ω representing heteroscedasticity. Substituting the $T \times 1$ vector t on the right side of (4.7) yields a $T \times 1$ vector $\hat{s}_{(tt)}$, as desired. Note that the initial T' components of $\hat{s}_{(tt)}$ are possibly negative and large (but not $-\infty$). The T'' outliers are possibly scattered anywhere in the sample. Still, we can replace the second column of the W_s matrix by $\exp(\hat{s}_{(tt)})$, and sort on the first column until it has elements $\tau = (1,2,...,T)'$, yielding a doubly sorted $T \times 2$ matrix denoted as W_{ss}, where the subscript "ss" suggests double sorting. Let the individual elements in the second column of W_{ss} be denoted by $W_{ss,t,2}$, which are time domain $\hat{\Omega}_{tt}$ quantities. Finally, our proposed correction for heteroscedasticity uses the transformation matrix:

$$\hat{V} = \text{diag}(1/W_{ss,t,2})^{1/2}. \tag{4.8}$$

Lemma 4.2. *The sorting map from the time domain to the values domain and the reverse map from the values domain to the time domain are linear (matrix) operations.*

Proof. In the absence of ties, the usual sorting yielding the order statistics $x_{(t)}$ from x_t is a function $S^O : \Re \to \Re$. It can be verified to be one-one onto (or bijection) implying that the reverse map must exist. Our maps t-dom ↔ v-dom carry along the time subscripts collected in the set $I^n = \{1,2,...,\}$ to facilitate the practical recovery of time subscripts in the reverse map, even if there are ties. To prove linearity, it suffices to show that these maps are matrix multiplications. Our constructive proof uses an example which does have two repeated values (ties). In this example, I^n has $\{1,2,3,4,5\}$ and $x_t = (4,8,36,20,8)$. The joint sorting reorders I^n as $I^n_s = \{1,2,5,4,3\}$ while $x_{(t)} = (4,8,8,20,36)$. Now we construct the mapping matrix for this example. Start with I_5 (identity matrix) and rearrange the diagonal ones to the positions given by I^n_s to create our O^r matrix. Verify that premultiplication of x_t by

$$O^r = \begin{bmatrix} 1\,0\,0\,0\,0 \\ 0\,1\,0\,0\,0 \\ 0\,0\,0\,0\,1 \\ 0\,0\,0\,1\,0 \\ 0\,0\,1\,0\,0 \end{bmatrix} \tag{4.9}$$

yields $x_{(t)}$. Also verify that premultiplying the order statistics $x_{(t)}$ by the inverse matrix $(O^r)^{-1}$ gives back the original x_t. Thus both maps in (t-dom

$\leftrightarrow$ v-dom) are linear operations for this example, despite the presence of a tie. Similar matrices can be constructed for any time series x_t. □

4.2.0.3 Cook–Weisberg Heteroscedasticity Testing

Now we turn to the Cook–Weisberg test using z_{ij} matrix of proxy data for heteroscedasticity. The $(j+1)$-st column of z_{ij} contains the j-th power of t for $j=0,1,\dots,h$. The true slope coefficients ϕ_j for $j=1,\dots,h$ of the polynomial of order h are all zero, under the null of homoscedasticity. The Cook–Weisberg model is

$$\Omega_{tt} = \exp(\sum_{j=0}^{h} \phi_j z_{ij}), \tag{4.10}$$

where z_{ij} is a $T \times (h+1)$ known matrix of arbitrary known quantities, which may or may not be related to one or more of the columns of X in their framework. White [31] considers a similar z_{ij} from "all second order products and cross products of original regressors." However, Greene [14, p. 509] criticizes that White's test is "nonconstructive," since on rejecting the null of homoscedasticity it fails to suggest a remedy. The algorithm proposed in this paper avoids such criticism. Conditional on our choice of the z_{ij} matrix, the Cook–Weisberg method tests the null of homoscedasticity. The validity of our choice of z_{ij} in (4.10) depends on our theorem proved in the sequel.

4.2.0.4 Analogy with Spectral Analysis

Our v-domain is somewhat analogous to the frequency domain of spectral analysis. Just as the cyclical properties are easier to study in the frequency domain, properties related to numerical magnitudes are easier to study in the v-domain. There are additional similarities. Priestley [20, p. 432] states that the variance of sample periodogram does not to tend to zero as $T \to \infty$, because it has "too many" sample autocovariances. Our incidental parameters problem is almost the same. Two Fourier integrals, [20, p. 201], allow two-way maps between the time and frequency domains. Our double sorting is simpler than Fourier integrals and allows similar two-way mappings between the time and v-domains.

The spectral kernel smoothers omit (down-weight) sample periodogram values outside a "window" to satisfy a technical assumption similar to our Assumption 4.3 below. We omit T'' additional "outliers" failing to satisfy the smoothness assumption. Similar to the following Theorem, Priestley [20, p. 464] proves consistency results in spectral analysis and notes the simplifying value of linearity.

Assumption 4.3 (truncation of s_{tt}). As sample size increases we omit a certain number of s_{tt} values and keep only T^{α} with $(0 < \alpha < 1)$, so that $(T^{\alpha}/T) \to 0$, as both T and $T^{\alpha} \to \infty$.

Since we are keeping only $T^{\alpha} = T - T' - T''$ observations, we can satisfy Assumption 4.3 provided $T' + T'' > 0$. Since we want $T'' > 0$, choosing a very high polynomial order h will make the fit too good and force the outlier detection constant M to be very small and threaten to make $T'' = 0$. Hence, I suggest a conservative choice of h.

Theorem 4.1. *Under Assumptions 4.1 to 4.3, the $W_{ss,t,2}$ in* (4.8) *yields consistent estimates of Ω_{tt} in the time domain, implying that $\Omega(\hat{\phi}) \Rightarrow \Omega(\phi)$.*

Proof. Recall the structure where a parameter point is a sequence of $\mathscr{F}(F)$ estimated by s_{tt}. Although the variance of each s_{tt} is $O(1/T)$, the variance of the vector (s_{tt}), denoted by parentheses, is $O(1)$. In the v-domain individual s_{tt} become the order statistics $s_{(tt)}$. There we omit $T' + T''$ observations to satisfy Assumption 4.3, so that $\text{var}(s_{(tt)}) \Rightarrow 0$ for the entire set. Substituting $s_{(tt)}$ in (4.7) and using Slutsky's theorem we have $\exp(\hat{s}_{(tt)}) \Rightarrow \Omega_{(tt)}$. Note that $W_{ss,t,2}$ provides a t-domain image of $\exp(\hat{s}_{(tt)})$, whereas Ω_{tt} is a t-domain image of $\Omega_{(tt)}$. Since the map from v-domain to time domain is linear (bijection) by Lemma 4.2, the structure in the sequences of $\mathscr{F}(F)$ remains intact and we have $W_{ss,t,2} \Rightarrow \Omega_{tt}$. □

Result 4.3 (Newey). Recall from (4.6) that we have in effect added a moment condition to estimate the ϕ vector in the values domain. By Lemma 4.2, this can be readily mapped back into the time domain, finishing the round trip through the v-domain promised in the introductory Sect.7.1. Following Newey [18] we accomplish asymptotic efficiency gain by adding the moment condition. Newey also notes the alternative condition $\mathrm{E}\varepsilon^3 \neq 0$ for efficiency gain.

So far we have established consistency and efficiency of our FGLS estimator. Next we turn to specification robustness by using two additional assumptions following Carroll and Ruppert [3].

Assumption 4.4. The alternative to $\Omega(\phi)$ of (4.1) is contiguous satisfying: $\Omega_{(tt)} = [1 + 2BT^{-1/2} f_t(X, \beta, \phi)]\Omega_{(tt)}$, where f_t is an arbitrary unknown function satisfying: $T^{-1}\Sigma_{t=1}^{T} f_t^2 \to \mu$, $(0 < \mu < \infty)$, and B is an arbitrary scalar. The FGLS uses preliminary OLS estimate b, satisfying $T^{1/2}(b - \beta) = O_p(1)$, to compute the residuals and $\hat{\phi}$ satisfies $T^{1/2}(\hat{\phi} - \phi) = O_p(1)$. Also suppose that errors ε in (4.1) are normally distributed and there is a positive definite matrix S_{pd}, such that we have $T^{-1}[X'\Omega^{-1}X] \Rightarrow S_{pd}$.

Assumption 4.5. Note that $\text{plim}_{T\to\infty} T^{-1}[X'\hat{\Omega}^{-1}X]$
$= \text{plim}_{T\to\infty} T^{-1}[X'\Omega^{-1}X]$, and $\text{plim}_{T\to\infty} T^{-1/2}[X'\hat{\Omega}^{-1}\varepsilon]$
$= \text{plim}_{T\to\infty} T^{-1/2}[X'\Omega^{-1}\varepsilon]$.

Theorem 4.2. *Assuming Assumption 4.1 to Assumption 4.5 the FGLS of (4.2) is efficient and robust. The asymptotic distribution of* $T^{-1/2}\,[b_{FGLS}-\beta]$ *is normal,* $N(0,S_{pd}^{-1})$*, under either the original* $\Omega(\phi)$ *or its contiguous alternative specifications.*

Proof. The efficiency of FGLS has been established in the literature, [14, Sect. 11.4], provided Assumption 4.5 holds and provided $\Omega(\hat{\phi}) \Rightarrow \Omega(\phi)$, which is proved in Theorem 4.1. Assumption 4.4 and a proof of robustness under contiguous (nearby) specification alternatives are in Carroll and Ruppert [3]. □

This completes our discussion of estimation of the square root of the inverse of $\hat{\Omega}$ matrix, denoted as V.

4.3 A Limited Monte Carlo Simulation of Efficiency of HE

Long and Ervin [16], Godfrey [13], Davidson and MacKinnon [7] and others have simulated the size and power of heteroscedasticity and autocorrelation consistent (HAC) estimators of SE(b). Cook and Weisberg's [4] simulation used cloud seeding data having $T=24$ observations. Let us inject objectivity in our Monte Carlo design by combining the Monte Carlo designs used by Long–Ervin with that of Cook–Weisberg. Of course, our focus is on efficient estimation of coefficients themselves, not inference. We pick x_1 to x_4 data from cloud seeding experiment (suitability criterion "sne," "cloudcover," "prewetness" and rainfall). These were among those chosen by Cook–Weisberg to allow wide variety of heteroscedasticity possibilities.

Our dependent variable is constructed artificially (as in both designs) by the relation

$$y = 1 + x_1 + x_2 + x_3 + x_4 + \varepsilon, \tag{4.11}$$

where all true coefficients including the intercept are set at unity, as in Long and Ervin (except that their coefficient of x_4 is zero) and where "ε" represents a vector of T random numbers chosen according to one of the following methods, which are called scedasticity functions by Long and Ervin. They have a far more extensive simulation and their focus is on HAC estimators. As $j=1,\ldots,J$ (=999) let $\varepsilon_{df5,j}$ denote a new vector of Student's t distributed (fat tails) independent pseudo-random numbers with five degrees of freedom.

SC1): $\varepsilon = \varepsilon_{df5,j}$. (no heteroscedasticity)
SC2): $\varepsilon = (x_1)^{1/2}\varepsilon_{df5,j}$. (disallows $x_1 < 0$)
SC3): $\varepsilon = (x_3+1.6)^{1/2}\varepsilon_{df5,j}$.
SC4): $\varepsilon = (x_3)^{1/2}(x_4+2.5)^{1/2}\varepsilon_{df5,j}$.
SC5): $\varepsilon = (x_1)^{1/2}(x_2+2.5)^{1/2}(x_3)^{1/2}\varepsilon_{df5,j}$.

We use HE estimators from (4.8) b'_1 to b'_4 (say) of the slopes in (4.11). The theory claims that GLS estimators are more efficient and that feasible GLS estimators are asymptotically more efficient than OLS under certain conditions. It is of interest to compare the efficiencies of OLS and GLS in a Monte Carlo simulation experiment with small T, to see if asymptotics holds for that T.

Our simulation program creates a four-dimensional array with dimensions (4, 6, 999, 5). The first dimension is for the estimates of $p = 4$ slope coefficients. The second dimension with six values is for the OLS, and five HE estimators denoted by HC0 to HC4 and described earlier in our discussion before Eq. (4.5). Of course, the "C" in HC refers to "correction" by our Eq. (4.8) not to the usual "consistency" in the sense used by Long and Ervin. The last dimension is for SC1 to SC5 in increasing order of heteroscedasticity severity.

After computing the standard deviations of 999 coefficient estimates we construct a summary array of dimension (4, 6, 5). It is convenient to suppress the first dimension and average the standard deviations over the four coefficients. Next, we divide the standard deviations for HC0 to HC4 by the standard deviation for OLS, reducing middle dimension to 5. The final results are reported in Fig. 4.1, where we look for numbers staying below the OLS vertical value of 1. The numbers 1 to 5 on the horizontal axis refer to HC0 to HC4. In this experiment, the sophisticated HC3 and HC4 corrections do not seem to offer great advantages in terms of efficiency. Since several values are below 1, many of our procedures are indeed more efficient than OLS. Not surprisingly, the efficiency improvement is generally higher when heteroscedasticity ought to be intuitively more severe (by looking at the complications of the formulas for SC1 to SC5 given above), although the intuition can fail in specific examples. In Fig. 4.1 the efficiency gain is the highest for the most severe SC5 (line marked "5") and lowest for the SC1 (marked "1") representing homoscedasticity, as might be expected. One can easily guard against SC1 by formal heteroscedasticity testing. Figure 4.2 is similar to Fig. 4.1, except that here we use $b'_4 = 0$ in (4.11) as in Long and Ervin. The efficiency gains over OLS continue to be achieved using correction formulas of HC0 to HC4 where all lines are below unity. In another experiment, we use (4.11) without the x_4 variable and economic data with nonmissing $T = 46$ observations from the "USfygt" data set of the "meboot" package. The x_1 to x_4 are: fygt1, infl, reallir, and usdef. Details are omitted to save space. Again, efficiency gains are clear except for SC1. Our experiments support the common practice of formally testing for heteroscedasticity before considering any corrections.

It is surprising that Long and Ervin's [16] large simulation finds that for typical sample sizes in economics ($T < 100$) the commonly used HC0, HC1 and HC2 methods provide inferior inference (in size and power) versus the simplest $s^2(X'X)^{-1}$ of OLS. In other words, OLS is hard to beat with $T < 100$. Yet we have chosen $T = 24, 46$ to raise the bar. Our HE method is worthy of study, since it is able to reduce the variance of OLS (over $J = 999$

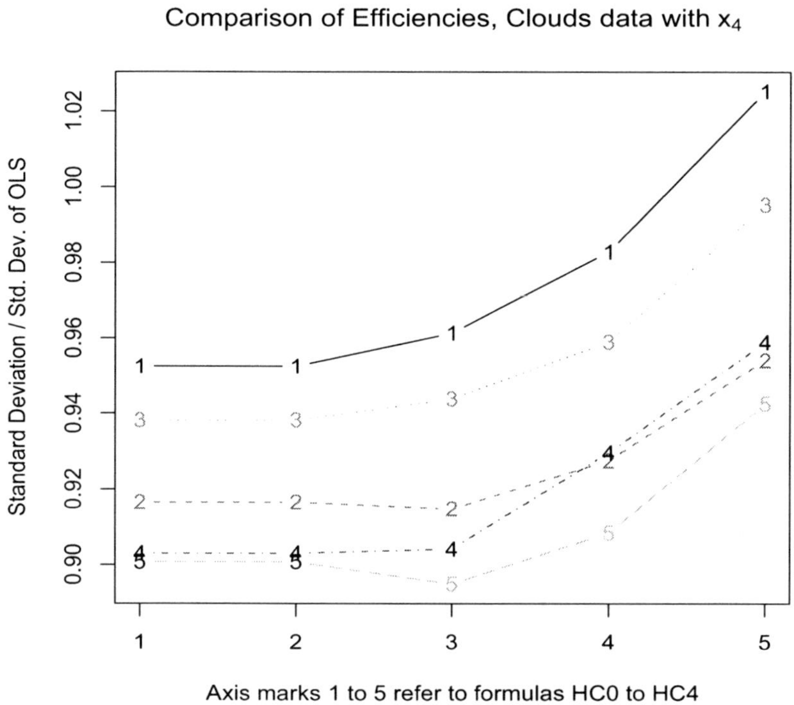

Fig. 4.1 SC1 to SC5 scedasticities built from Student's t (df=5) with lines marked 1 to 5: Results of 999 experiments.

experimental values) when heteroscedasticity is present in the model. Cook and Weisberg emphasize a need to supplement simulations with graphics. Figures 4.1 to 4.2 illustrate some interesting patterns of heteroscedasticity in econometric applications.

So far we have considered efficient estimation of β in (4.1). It is known in the literature that the model may be subject to endogeneity and identification problems, which primarily affect statistical inference. These problems, especially the latter are best described in the context of an example. Hence let us postpone discussion of superior inference until after the next section.

4.4 An Example of Heteroscedasticity Correction

This section illustrates our heteroscedasticity correction (4.8) with a model having three regressors. We use the augmented Phillips curve model by Alogoskoufis and Smith [1] using annual time series from 1857 to 1987 from

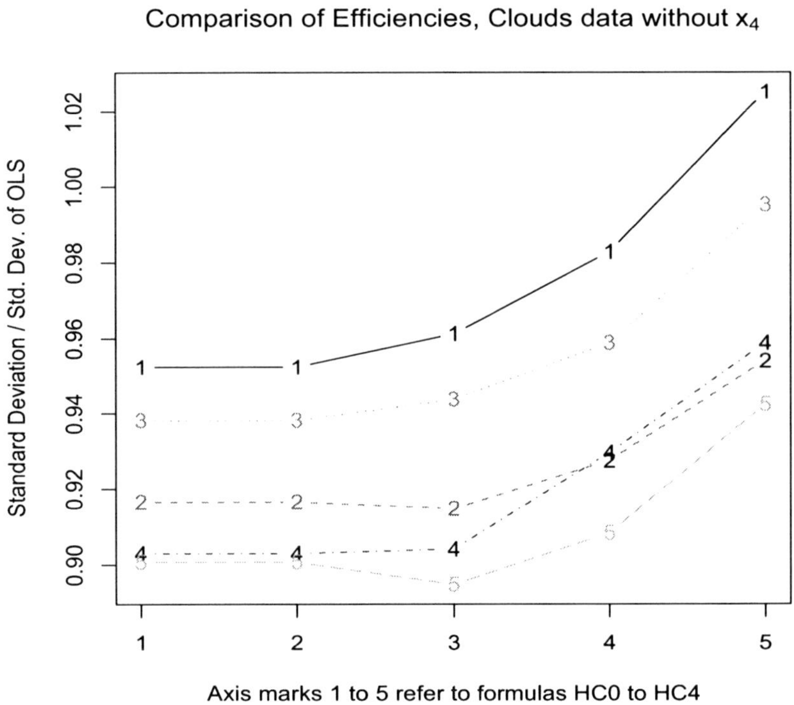

Fig. 4.2 SC1 to SC5 scedasticities built from Student's t (df=5) with lines marked 1 to 5: Results of 999 experiments.

the United Kingdom (U.K.) along with an expected inflation variable. The Phillips curve empirically measures the following assertion: the lower the unemployment in an economy, the higher the rate of increase in nominal wages. Write a loglinear, expectations-augmented version of Phillips' wage equation as

$$\Delta w_t = \alpha_0 + \alpha_1 E(\Delta p_{c,t} | I_{t-1}) + \alpha_2 \Delta u_t + \alpha_3 u_{t-1} + \varepsilon_t, \tag{4.12}$$

where Δ is the difference operator, t is the time subscript, w is the log of nominal wages, p_c is the log of consumer prices, u is the unemployment rate, and ε is a white noise error term. The α's are parameters, E is the mathematical expectations operator, and I_{t-1} is the information set of wage-setters at time $t-1$.

If we ignore the expectation operator E, we face possible endogeneity of the inflation variable, leading to possible inconsistency of OLS. Alogoskoufis and Smith [1] show that the following AR(1) process for rational expectation of price inflation can solve endogeneity leading to the model:

$$E(\Delta p_{c,t}|I_{t-1}) = \pi(1-\rho) + \rho \Delta p_{c,t-1}, \quad (4.13)$$

where π is the steady-state inflation rate, and ρ is the autoregressive coefficient.

Substituting (4.13) in (4.12),

$$\Delta w_t = \alpha_0' + \alpha_1 \rho \Delta p_{c,t-1} + \alpha_2 \Delta u_t + \alpha_3 u_{t-1} + \varepsilon_t, \quad (4.14)$$

where $\alpha_0' = \alpha_0 + \alpha_1 \pi(1-\rho)$. Thus these authors have a system of two equations (4.13) and (4.14). This paper focuses on estimation and inference for (4.14). The estimation will correct for its statistically significant heteroscedasticity.

Note that the coefficient α_1 in (4.12) has now become $\alpha_1 \rho$ in (4.14). Thus we have a new *identification* problem due to what Dufour [9] calls "locally almost unidentifiable" parameters and hence we must contend with his "impossibility theorems" showing that Wald-type inference (t test) is "fundamentally flawed" due to unbounded confidence sets and zero coverage probability. Instead of Fisher's pivot used in Wald-type t tests, we use Godambe's pivot in the next section.

Table 4.1 Phillips curve heteroscedasticity tests and efficient estimation

Transform of $\hat{u}^2$	Order of τ^h polynomial	p-value Cook Weisberg	Adjusted R^2	Coef. $\hat{\alpha}_2$ of Δu_t after V	t-stat for $\hat{\alpha}_2$
1	2	3	4	5	6
HC0 & HC1*	linear	9.031e-09	0.8539608	-2.7174027	-4.541935
HC2	linear	1.2427e-08	0.8483255	-2.9410992	-4.6847318
HC3	linear	2.074e-08	0.8428199	-3.1837852	-4.8397987
HC4	linear	3.3201e-08	0.8289122	-3.1257282	-4.2312002
HC0& HC1	quadratic	6.6322239e-05	0.9448547	-4.0058258	-6.5646845
HC2	quadratic	5.3017918e-05	0.934911	-4.2014699	-6.4314288
HC3	quadratic	4.6128875e-05	0.9263111	-4.4148725	-6.3677521
HC4	quadratic	4.6484793e-05	0.9106575	-4.5775178	-5.6541541
HC0& HC1	cubic	4.6975e-08	0.9896981	-4.7653311	-6.5022282
HC2	cubic	5.9871e-08	0.98676	-5.1609628	-6.1181163
HC3	cubic	1.10414e-07	0.9882737	-5.5840857	-5.9284502
HC4	cubic	2.85749e-07	0.9895367	-6.1668222	-5.6894456
HC0 & HC1	quartic	3.31095e-07	0.9904292	-4.9242539	-6.3482746
HC2	quartic	4.62814e-07	0.9884053	-5.4626662	-5.9190534
HC3	quartic	6.92653e-07	0.9906989	-6.0132562	-5.7528265
HC4*	quartic	1.051275e-06	0.9919242	-6.687974	-5.6190831

Our OLS estimation of (4.14) is reported in Table 4.2 where the coefficient α_2 of Δu_t is seen to be negative, but statistically insignificant with a low t-value and high p-value. This suggests that the trade-off between wages and unemployment claimed by Phillips is not statistically significant in these

Table 4.2 Heteroscedasticity efficient (HE) feasible GLS estimates of coefficients of Phillips curve, HC4 weights and quartic polynomial

Variable	Estimate	Std. Error	t value	Pr(> | t |)
OLS Estimates				
(Intercept)	0.0387	0.0096	4.02 (4.5)	0.0003
$p_{c,t-1}$	0.7848	0.1396	5.62 (4.5)	0.0000
Δu_t	-0.8457	0.8465	-1.00 (-1.35)	0.3244
u_{t-1}	0.0147	0.1451	0.10 (0.16)	0.9197
GLS Estimates				
(Intercept)	0.0057	0.0321	0.18	0.86
$p_{c,t-1}$	1.7894	0.1665	10.75	8.7e-13
Δu_t	-6.6878	1.1902	-5.62	2.2e-06
u_{t-1}	0.0110	0.1182	0.09	0.93
Confidence intervals are given below				
limits →	2.5 %	97.5 %		
(Intercept)	0.019191	0.058198		
$p_{c,t-1}$	1.45177	2.12710		
Δu_t	-9.10186	-4.27409		
u_{t-1}	-0.22876	0.25082		

Notes: Residual standard error: 0.169 on 36 degrees of freedom, F(3, 36): 92.5, p-value: $< 2e-16$. In the column for t-values under OLS we report in parentheses the t-values based on heteroscedasticity corrected standard errors.

data. If we use the heteroscedasticity corrected (HC) standard errors the t-values change somewhat (reported in parentheses in Table 4.2) but remain insignificant.

Upon fitting (4.14) to data, none of the OLS residuals is zero ($T' = 0$ here). The residual autocorrelations are not statistically significant. The p-values for the Breusch–Godfrey test for serial correlation of orders 1 through 4 are respectively (0.06691, 0.07763, 0.1511, and 0.09808) suggesting nonrejection of the null hypothesis of zero autocorrelation among regression errors at the 5% level. R software tools useful in correcting for autocorrelation from fairly general ARMA(p,q) process are discussed in the Appendix.

We find that heteroscedasticity is a problem for this example. Studentized Breusch–Pagan test yields the statistic = 12.69, $df = 3$, p-value = 0.005362, implying rejection of the assumption of homoscedasticity.

Table 4.1 lists key results for 20 choices of HCj for $j = 0, \ldots, 4$ and τ^h for $h = 1, \ldots, 4$. The first two columns identify the HCj and τ^h. Column 3 has p-values for the Cook–Weisberg score test for the presence of heteroscedasticity, based on our new choice for their artificial matrix of z_{ij}. The $(j+1)$-st column of z_{ij} contains the j-th power of sorted (τ_s) for $j = 0, 1, \ldots, h$. If there is no heteroscedasticity, all the coefficients ϕ in (4.10) will be insignificant. Since the p-values in column 3 are near zero, homoscedasticity is rejected for all

20 choices. We expect researchers to use formal tests for heteroscedasticity before deciding to correct for it.

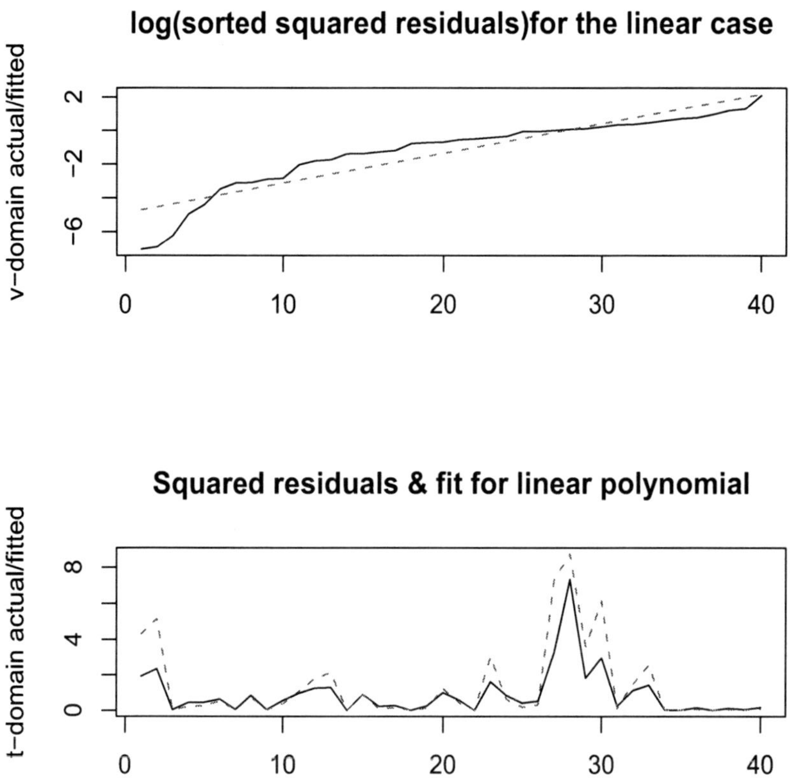

Fig. 4.3 Phillips curve HC1 & linear case. **a** log(sorted squared residuals) for the linear case. **b** Squared residuals & fit for linear polynomial.

Column 4 of Table 4.1 has the adjusted R^2 for the regression of sorted logs of $(\hat{u})^2$ on powers of sets of integers $\{t\}$ in (4.4). Column 5 has estimates $\hat{\alpha}_2$ from (4.14). Column 6 has the t-statistic for $\hat{\alpha}_2$ associated with the variable Δu_t after the premultiplication by $\hat{V}$ based on the particular choice of HCj and polynomial power h. We choose two rows marked with an $(*)$ for further analysis: HC1 with linear polynomial with the lowest t-statistic on α_2 and HC4 with quartic $(h = 4)$. The starred choices have adjusted R^2 =0.854 and 0.992, respectively.

Figure 4.3 reports graphs for the linear HC1 case. The upper panel plots the order statistics of logs of squared residuals $s_{u,0}$ as the solid line along with a dashed line for the fitted values from a simple straight line in time, representing our smooth intermediate function g. After computing $\exp(g)$ and using the reverse map, we get the second column of doubly sorted W_{ss} matrix. The lower panel plots them in the time domain as the dashed line along with the solid line representing original squared residuals over time. It is clear from both figures that with only two parameters of ϕ (intercept and coefficients of τ) we are able to get good estimates of heteroscedastic variances, thanks to the double sort.

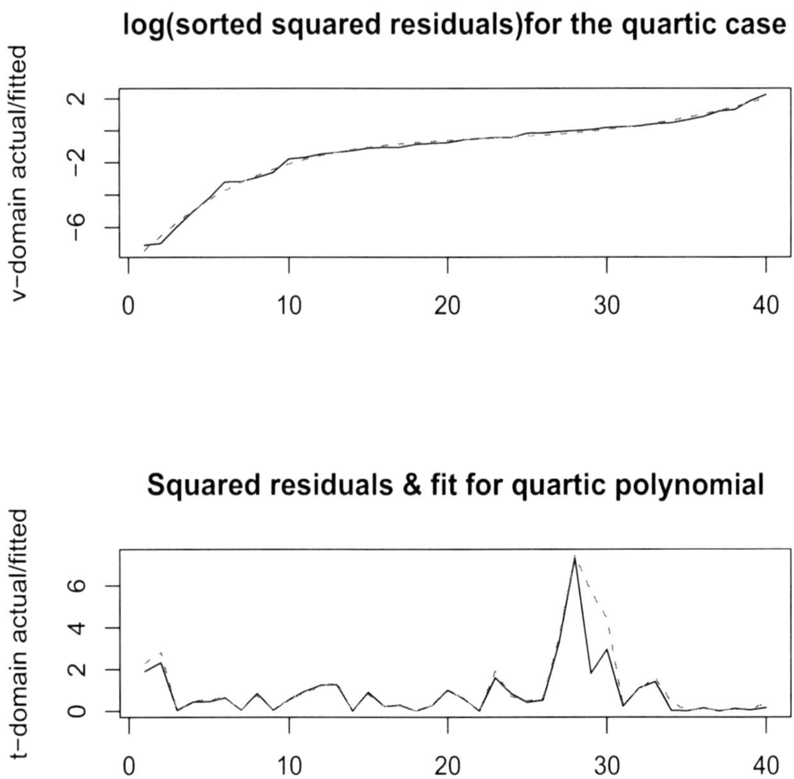

Fig. 4.4 Phillips curve HC4 & quartic case. **a** log(sorted squared residuals) for the quartic case. **b** Squared residuals & fit for quartic polynomial.

Figure 4.4 is similar to Figure 4.3, except that we have the quartic case with HC4 here, showing that the quartic fit is better. After all, HC4 with the quartic has a high (=0.992) adjusted R^2 in Table 4.1. The figures show that rearranging observations in an increasing order of squared residuals can reveal hidden heteroscedasticity with just a few additional parameters in ϕ. The original OLS, as well as all 20 cases of efficient estimation show significantly negative α_2, implying support for the Phillps curve trade-off between unemployment and wages.

Table 4.2 reports details for the row HC4* of Table 4.1. It has feasible GLS estimates after heteroscedasticity correction by a quartic under HC4 transformation of squared residuals. The F(3, 36) statistic for the overall fit is 92.5 with a near-zero p-value. The OLS coefficient of Δu_t which was statistically insignificant for OLS before heteroscedasticity correction, has now become significantly negative at the 5% level.

4.5 Superior Inference of Deep Parameters Beyond Efficient Estimation

Now we are ready to consider improved inference for (4.14) upon recognizing that a deep parameter of interest from Phillips' model (4.12) might be "locally almost unidentifiable." Following Vinod [29, Sect. 10.4] we now use Godambe's [12] pivot function (GPF) relying on his theory of estimating functions discussed in Vinod [29, Sect. 10.3]. The GPF is defined as

$$\mathrm{GPF} = \sum_{t=1}^{T} g_t^* \Big/ \left[\sum_{t=1}^{T} E(g_t^*)^2\right]^{1/2}, \tag{4.15}$$

where g_t^* is the "scaled quasi-score function" from the underlying quasi-likelihood function also known as the optimal estimating equation.

This pivot avoids the Wald-type pivotal statistic commonly used in the usual t-tests. Vinod [26] extends the GPF to the multivariate regression problems. In the simpler scalar case he rewrites the GPF as a sum of T scaled quasi-scores:

$$\mathrm{GPF} = \sum_{t=1}^{T} S_t/S_c = \sum_{t=1}^{T} \tilde{S}_t, \quad \text{where } S_c = \left[\sum_{t=1}^{T} E(S_t)^2\right]^{1/2}, \tag{4.16}$$

where we denote scaled quasi-score functions as: $\tilde{S}_t$. As a sum of T items, the central limit theorem assures us that GPF $\sim N(0,1)$ is asymptotically unit normal. Thus, the probability distribution of GPF is independent of unknown parameters and therefore it is a pivot.

The asymptotic normality of GPF means that marginal densities for individual elements of $\boldsymbol{\beta}$ in the notation of (4.1) are also Normal, allowing us to construct 95% confidence intervals (CI95). Following Vinod [29, Sects. 10.4.1 and 10.4.2] we apply the R function called "gpf" provided in a snippet. It uses a scalar version of GPF by using the Frisch–Waugh theorem, and constructs CI95 for all regression coefficients in a sequence.

Table 4.3 Confidence intervals for 3 slope coefficients after GPF

coef.	obs.val	Lower	Upper
$\hat{\alpha}_1$	1.79	1.69	2.14
$\hat{\alpha}_2$	-6.69	-11.44	-6.25
$\hat{\alpha}_3$	0.01	-0.06	0.06

Note that the confidence interval for α_2 does not contain the zero. Thus, even though the OLS coefficient of Δu_t was statistically insignificant for OLS before heteroscedasticity correction, it has now become significantly negative at the 5% level after the correction and the inference based on the superior pivot function (GPF) continues to support that it is significantly negative, consistent with Phillips' notion of a trade-off.

4.6 Summary and Final Remarks

This paper suggests some practical solutions to the problem of heteroscedastic errors. If squared residuals $(\hat{u})^2$ are nonconstant, we suggest making them monotonic by sorting, and finding their fitted values by regressing $\log(\hat{u})^2$ on powers of a suitable subset from the set of integers $(1, 2, \ldots, T)$. A (t-dom $\leftarrow$ v-dom) map involving double sorting recovers the original time subscript for $(\hat{u})^2$, fitted with very few additional parameters in ϕ and results in a new, practical and parsimonious correction for heteroscedasticity. Figures 4.3 and 4.4 illustrate this rather well for the example of Phillips curve using U.K. data, where heteroscedasticity problem is present. We find that efficient estimation converts an insignificant trade-off between wages and unemployment into a significant one.

We have considered the possibility that parameters are "locally almost unidentified" suggested by the impossibility theorems proved by Dufour [9]. Hence we abandon the problematic Fisher pivot of the usual (Wald-type) t tests in favor of the Godambe pivot function (GPF) which is always a sum of T quantities (scaled scores) and hence asymptotically normal by the central limit theorem. We confirm a trade-off relation between wages and unemployment (significantly negative slope coefficient) with the use of the superior inference based on the GPF.

We discuss how mappings between the time domain and the new (ordered numerical values) v-domain are analogous to mappings in spectral analysis between time and frequency domains (through Fourier integrals). Assuming smoothness of heteroscedasticity in the v-domain, and two further assumptions, a theorem proves consistency of our HE estimator. A simulation experiment uses a published design where OLS is found hard to beat in samples with $T < 100$. We still use small samples ($T = 24, 46$) and report efficiency gains over OLS achieved by our new HE estimators. The simulation also supports the common practice of formally testing for heteroscedasticity before considering any corrections.

Appendix: Efficient Estimation Under ARMA(p, q) Errors

If regression errors are autocorrelated, the off-diagonal elements of Ω are nonzero. For example, if the vector of errors in equation (4.1) denoted by ε, follows a first-order autoregressive process, AR(1): $\varepsilon_t = \rho\varepsilon_{t-1} + \zeta_t$, then the error covariance matrix for ε is

$$\mathrm{Cov}(\varepsilon) = \Omega = \sigma^2(1-\rho^2)^{-1}\begin{bmatrix} 1 & \rho & \rho^2 & \cdots & \rho^{T-1} \\ \rho & 1 & \rho & \cdots & \rho^{T-2} \\ \vdots & \vdots & \vdots & \vdots & \vdots \\ \rho^{T-1} & \rho^{T-2} & \rho^{T-3} & \cdots & 1 \end{bmatrix}. \tag{4.17}$$

This large matrix is completely known when estimates of ρ and σ^2 are available. A general way of thinking about this is proposed in Vinod [25, 29, Chap. 2] where it is argued that the *order of dynamics* of any ARMA $(p, p-1)$ process is dictated by the underlying stochastic difference equation of order p. Here we consider the case of regression errors satisfying stationary and invertible ARMA(p,q) models. When the errors are nonstationary, they are revealed by unit root testing. Our corrections are not suitable for the nonstationary case.

Our focus is on using some powerful numerical methods available in R without attempting analytical expressions for Ω or Ω^{-1} which rely on theoretical ingenuity and which are discussed in many textbooks including Vinod [29]. The key theoretical result is that the Ω matrix is a Toeplitz matrix. Given a vector of some numbers (e.g., 1 to 4) a typical Toeplitz matrix is given in R by the function "toeplitz." For example, the R command "`toeplitz(1:4)`" produces the following:

$$\begin{bmatrix} 1 & 2 & 3 & 4 \\ 2 & 1 & 2 & 3 \\ 3 & 2 & 1 & 2 \\ 4 & 3 & 2 & 1 \end{bmatrix} \tag{4.18}$$

In the realm of numerical methods if we have a suitable vector v of autocorrelation coefficients we can readily construct our Ω matrix as a Toeplitz matrix created by the R command "toeplitz(v)." This means that we simply need to estimate the autocorrelation function (acf) for the suitable ARMA(p,q) model representing the regression error process.

Thus we consider a two-step process, where the first step is computation of OLS residuals r_t. Next, we fit several ARMA(p,q) models, where $p,q = 1,2,3,4$ with $q \le p$ to the residuals, keeping track of the Akaike Information Criterion (AIC). Next we sort the models according to the AIC, where the minimum AIC model will be at the top. In the interest of pasimony we view the AIC values and choose the most parsimonious model by minimizing the number of parameters $p+q$. Upon choosing the ARMA(p,q) model, the R package called "fArma" provides a function called "ARMAacf" to compute the vector v and then uses it to construct the feasible and parsimonious Ω matrix.

Once the diagonal matrix Λ of T eigenvalues and the large $T \times T$ matrix Z of eigenvectors are known, the inverse of the Ω matrix is readily written numerically and substituted in the GLS estimator. As before, we can construct $V = \hat{\Omega}^{-1/2}$ the square root matrix of the inverse of $\hat{\Omega}$ by writing

$$\hat{\Omega}^{-1/2} = Z\hat{\Lambda}^{-1/2}Z'. \tag{4.19}$$

Now we simply substitute this in (4.3) and obtain efficient estimates of regression coefficients despite autocorrelation among regression errors of fairly general type, ARMA(p, q). As a practical matter we simply have to regress Vy on VX and use the "meboot" and GPF to obtain superior confidence intervals on efficient estimates of regression coefficients.

As a numerical example we use an abridged version of the Phillips model (4.14) such that it does have autocorrelated errors. We use

$$\Delta w_t = \alpha_0'' + \alpha_2'' \Delta u_t + \varepsilon_t. \tag{4.20}$$

OLS estimation of the above model yields the following results:

Table 4.4 OLS Estimation Results

	Estimate	Std. Error	t value	Pr(>|t|)
$\hat{\alpha}_0''$	0.0817	0.0069	11.82	0.0000
$\hat{\alpha}_2''$	2.2051	0.8923	2.47	0.0181

Table 4.5 Durbin-Watson statistics

Lag	r=auto-correlation	DW statistic	p-value
1	0.61	0.61	0.00
2	0.45	0.45	0.00
3	0.35	0.35	0.04
4	0.47	0.47	0.00

The p-values for the Durbin–Watson test statistic for the first four lag orders are smaller than the usual type I error of 0.05 implying that regression errors for (4.20) are autocorrelated. The 95% OLS confidence interval for the slope is (0.39882, 4.01144). We use the residuals of this first stage regression to compute ARMA(p,q) estimates and their AIC values.

Table 4.6 ARMA(p,q) Summary Results

p	q	AIC
4	4	-158.60
4	3	-157.62
4	2	-157.23
3	3	-157.11
1	1	-156.83
3	2	-156.25
2	1	-155.11
2	2	-154.91
4	1	-154.78
3	1	-152.41

The lowest AIC is for ARMA(4,4) at -158.6 and next is ARMA(4,3) at -157.6. Perhaps the most parsimonious error specification, ARMA(1,1), has the fifth lowest AIC of -156.8, which is not much larger than the lowest AIC.

The ARMA(4,3) has $\hat{\sigma}^2$=0.000665 and ARMA(1,1) has $\hat{\sigma}^2$=0.0009859. The coefficients and respective standard errors are listed below:

```
ARMA(4,3) coefficients and standard errors.
The log likelihood equals 86.81 and AIC= -157.6
        ar1     ar2     ar3    ar4     ma1    ma2    ma3
      0.919  -0.519  -0.133  0.458  -0.022  0.003  0.391
s.e.  0.298   0.423   0.393  0.211   0.309  0.333  0.311

ARMA(1,1) coefficients and standard errors.
The log likelihood equals  81.42 and AIC= -156.83
          ar1      ma1
       0.8119  -0.3534
s.e.   0.1507   0.2664
```

We input these estimates into an R function to compute the square root of the inverse of Ω matrix as new V matrix and estimate the regression of appropriately defined Vy on VX. Finally, we use the GPF with maximum entropy bootstrap having $J = 999$ replications and find 95% GPF confidence intervals. Note that the OLS interval CI95=(0.39882, 4.01144) is 5.2 times wider than CI95=(1.334, 2.028) using ARMA(4,3) error correction. OLS CI95 remains 4.49 times wider than CI95=(0.555, 1.359) using a parsimonious ARMA(1,1) error correction, further confirming that we have achieved efficiency gains.

References

1. Alogoskoufis, G., Smith, R.: The Phillips curve, the persistence of inflation, and the Lucas critique: Evidence from exchange rate regimes. American Economic Review **81**(2), 1254–1275 (1991)
2. Carroll, R.J.: Adapting for heteroscedasticity in linear models. Annals of Statistics **10**, 1224–1233 (1982)
3. Carroll, R.J., Ruppert, D.: A comparison between maximum likelihood and generalized least squares in a heteroscedastic linear model. Journal of the American Statistical Association **77**, 878–882 (1982)
4. Cook, D., Weisberg, S.: Diagnostics for heteroscedasticity in regression. Biometrika **70**, 1–10 (1983)
5. Cragg, J.G.: More efficient estimation in the presence of heteroscedasticity of unknown form. Econometrica **51**(3), 751–763 (1983)
6. Cribari-Neto, F.: Asymptotic inference under heteroskedasticity of unknown form. Computational Statistics & Data Analysis **45**, 215–233 (2004)
7. Davidson, R., MacKinnon, J.G.: Econometric Theory and Methods. Oxford University Press, New York (2004)
8. Dixon, S.L., McKean, J.W.: Rank-based analysis of the heteroscedastic linear model. Journal of the American Statistical Association **91**, 699–712 (1996)
9. Dufour, J.M.: Some impossibility theorems in econometrics with applications to structural and dynamic models. Econometrica **65**, 1365–1387 (1997)
10. Eicker, F.: Asymptotic normality and consistency of the least squares estimator for families of linear regressions. Annals of Mathematical Statistics **34**, 447–456 (1963)
11. Fox, J.: car: Companion to Applied Regression. R package version 1.2-14 (2009). URL `http://CRAN.R-project.org/package=car`. I am grateful to Douglas Bates, David Firth, Michael Friendly, Gregor Gorjanc, Spencer Graves, Richard Heiberger, Georges Monette, Henric Nilsson, Derek Ogle, Brian Ripley, Sanford Weisberg, and Achim Zeileis for various suggestions and contributions
12. Godambe, V.P.: The foundations of finite sample estimation in stochastic processes. Biometrika **72**, 419–428 (1985)
13. Godfrey, L.G.: Tests for regression models with heteroskedasticity of unknown form. Computational Statistics & Data Analysis **50**, 2715–2733 (2006)
14. Greene, W.H.: Econometric Analysis, 4 edn. Prentice Hall, New York (2000)
15. Kendall, M.G., Stuart, A.: The Advanced Theory of Statistics, vol. 2. Macmillan Publishing, New York (1979)
16. Long, J.S., Ervin, L.H.: Using heteroscedasticity consistent standard errors in the linear regression model. The American Statistician **54**, 217–224 (2000)
17. McCullough, B.D., Vinod, H.D.: Verifying the solution from a nonlinear solver: A case study. American Economic Review **93**(3), 873–892 (2003)

18. Newey, W.K.: Efficient estimation of models with conditional moment restrictions. In: G.S. Maddala, C.R. Rao, H.D. Vinod (eds.) Handbook of Statistics: Econometrics, vol. 11, chap. 16. North–Holland, Elsevier Science Publishers, New York (1993)
19. Pregibon, D.: Data analysts captivated by r's power. The New York Times **January 6** (2009). URL `http://www.nytimes.com/2009/01/07/technology/business-computing/07program.html?_r=1\&em`
20. Priestley, M.B.: Spectral Analysis and Time Series, vol. I and II. Academic Press, London (1981)
21. R Development Core Team: R: A Language and Environment for Statistical Computing. R Foundation for Statistical Computing, Vienna, Austria (2008). URL `http://www.R-project.org`. Cited 14 May 2009
22. R Development Core Team: R: A language and environment for statistical computing. R Foundation for Statistical Computing, Vienna, Austria (2009). URL `http://www.R-project.org`. ISBN 3-900051-07-0
23. Rao, P.S.R.S.: Theory of the MINQUE— A Review. Sankhya **39**, 201–210 (1977)
24. Robinson, P.M.: Asymptotically efficient estimation in the presence of heteroskedasticity of unknown form. Econometrica **55**(4), 875–891 (1987)
25. Vinod, H.D.: Exact maximum likelihood regression estimation with arma $(n,n\text{-}l)$ errors. Economics Letters **17**, 355–358 (1985)
26. Vinod, H.D.: Foundations of statistical inference based on numerical roots of robust pivot functions (fellow's corner). Journal of Econometrics **86**, 387–396 (1998)
27. Vinod, H.D.: Ranking mutual funds using unconventional utility theory and stochastic dominance. Journal of Empirical Finance **11**(3), 353–377 (2004)
28. Vinod, H.D.: Maximum entropy ensembles for time series inference in economics. Journal of Asian Economics **17**(6), 955–978 (2006)
29. Vinod, H.D.: Hands-on Intermediate Econometrics Using R: Templates for Extending Dozens of Practical Examples. World Scientific Publishers, Hackensack, NJ (2008). URL `http://www.worldscibooks.com/economics/6895.html`. Cited 14 May 2009.
30. Vinod, H. D. and Lopez-de-Lacalle, Javier: Maximum entropy bootstrap for time series: The meboot R-package. Journal of Statistical Software **29**(5), 1–30 (2009). URL `http://www.jstatsoft.org`. Cited 14 May 2009.
31. White, H.: A heteroskedasticity-consistent covariance matrix and a direct test for heteroskedasticity. Econometrica **48**, 817–838 (1980)

Chapter 5
Bubble Plots as a Model-Free Graphical Tool for Continuous Variables

Keith A. Markus and Wen Gu

Abstract Researchers often wish to understand the relationship between two continuous predictors and a common continuous outcome. Many options for graphing such relationships, including conditional regression lines or 3D regression surfaces, depend on an underlying model of the data. The veridicality of the graph depends upon the veridicality of the model, and poor models can result in misleading graphs. An enhanced 2D scatter plot or *bubble plot* that represents values of a variable using the size of the plotted circles offers a model-free alternative. The R function *bp3way()* implements the bubble plot with a variety of user specifiable parameters. An empirical study demonstrates the comparability of bubble plots to other model-free plots for exploring three-way continuous data.

5.1 Introduction

It is easy to overlook descriptive statistics as a means to better understand data. Although usually inadequate as a complete analysis, descriptive graphs such as histograms, box-and-whisker plots, stem-and-leaf plots, and scatter plots can quickly and directly convey information about data distributions, outliers, skewness, kurtosis, and floor and ceiling effects at a glance. They are useful data screening tools that provide researchers a better understanding of results expressed in numerical summaries [12, 23]. Moreover, as social science research moves beyond looking only at main effects to explain complex

Keith A. Markus
John Jay College of Criminal Justice of The City University of New York, New York, NY 10019, USA e-mail: kmarkus@aol.com

Wen Gu
John Jay College of Criminal Justice of The City University of New York, New York, NY 10019, USA e-mail: wgu@gc.cuny.edu

H.D. Vinod (ed.), *Advances in Social Science Research Using R*, Lecture Notes in Statistics 196, DOI 10.1007/978-1-4419-1764-5_5,

behaviors, detection of interactions for finding moderators and conducting meditational studies become increasingly important for drawing conclusions about causal relationships [15, 2, 13]. Thus, graphs that depict interactions may allow researchers a quick check of their data to determine the need for further analyses. Well-designed graphs are more revealing and convincing, because they can clarify data and maintain an audience's attention [21]. Methods for graphing interactions can also help researchers make better use of statistical models and report their results more effectively.

This chapter considers options for graphing the joint distribution of three continuous variables, typically one outcome and two predictors, that require minimal assumptions about the relationship or the underlying generating model. The initial portion of this chapter considers several alternatives for plotting three continuous variables and reviews the relevant literature. The middle portion describes an R function for plotting such relationships in as a three-way bubble plot, essentially an enhanced two-way scatter plot with the size of the bubbles proportional to the value of the third variable. The remainder of the chapter reports an empirical study comparing three promising plots using a variety of relationships between the outcome variable and the two predictors, and offers tentative conclusions based on this research.

5.2 General Principles Bearing on Three-Way Graphs

Properly used, graphs can be powerful tools for depicting information. They can convey complex ideas with clarity, precision, and efficiency [22]. However, only well-designed graphs will serve this purpose [21] effectively. Effective use of graphs is what Tufte [22] termed "graphical excellence" defined as "what gives to the viewer the greatest number of ideas in the shortest amount of time with the least ink in the smallest space" (p. 51). Through empirical studies and better understanding of the human visual system, statisticians and researchers have developed explicit guidelines for achieving graphical excellence. According to Cleveland [5] and Robbins [19], internal factors affecting perceptual accuracy and detection include (in decreasing order of ease): position along a common scale, position along identical scale, length, angle, area, volume, and color. Many external factors affect effective graphing, such as audience skills, the purpose of the graph, and the complexity of information. Schmid [21] recommended that good graphs are accurate, simple, clear, appealing, and well structured.

When graphs show more than two variables on two-dimensional surfaces, as when printed on paper, graph designers must find ways to incorporate additional variables without confusing the reader. Presenting multivariate data is more difficult than presenting univariate or bivariate data because the media on which graphs are drawn (paper, computer displays) are two-dimensional [10]. Although graphs displayed on computer screens can be rotated to better conceptualize the third dimension for additional variables, publication generally requires a static, two-dimensional image. When data move to even three variables and three dimensions, observers must infer the multidimensional structure from a two-dimensional medium [5]. Inferring multiple dimensions, and consequently reading graphs, heavily depends on the human visual system.

During the 1960s and 1970s, cartographers, psychophysicists, and psychologists sought to create better methods for displaying multivariate data. Some new methods included: Chernoff faces, Anderson metroglyphs, Cleveland–Kleiner weathervane plots, Diaconis–Friedman M and N plots, Tukey–Tukey dodecahedral views, Kleiner–Hartigan trees, Andrews curves, Tufte rug plots, and the scatter plot matrix [5]. Only the scatter plot matrix had any success and is still commonly used. Cleveland [5] argued that not enough attention was paid to graphical perception, which involves a better understanding of how the human visual system decodes and encodes data. Robbins [19] concurred, stating that creating more effective graphs involves choosing a graphical construction that is visually decoded easily on the ordered list of elementary graphical tasks, while balancing this ordering with consideration of distance and detection.

As psychologists understood human perception better, they realized that spatial perception, especially depth perception, is heavily influenced by learning and past experience. To maintain perceptual constancies of color, size, and shape, the visual system is susceptible to various systemic distortions in spatial perception, commonly known as optical illusions [20]. One such illusion occurs when humans perceive depth in a two-dimensional surface. A frequently used method to create the illusion of depth is tilting or rotating the already established horizontal and vertical axes, thereby creating an appearance of the third dimension that can be used for the third variable. This is problematic during graph interpretation, however, because the perception of the additional dimension is achieved by tricking the perceptual system. Cubes drawn on paper often produce a Necker illusion, in which one can see two cubes interchangeably from the same drawing by focusing on different parts of the cube as the foreground. The Necker illusion occurs as a result of the Moiré effect, in which graph design interacts with the physiological tremor of the eye to produce the distracting appearance of vibration or movement [22], and the Müller–Lyer illusion, in which two physically equal lines can be viewed as having different lengths depending on the directions of arrows added to the lines [20]. The Moiré effect gives the observer the appearance of seeing two cubes while the Müller–Lyer illusion aids this misperception by using ambiguous cube corners as distance cues [20]. Thus, the instability of human perception to anchor lines for distance cues makes cube-like graphs difficult to perceive and interpret, and three-dimensional graphs are typically difficult to interpret on two-dimensional surfaces.

Perceiving depth on two-dimensional surfaces involves simulated perspective projection, which often distorts perception [21], thus statisticians and graph designers generally warn against using three-dimensional plots. For example, bar graphs with receding depth give the appearance that bar heights are higher than their actual heights. Graphs designers are opposed to using more dimensions than the number of information-carrying variables when graphing two or fewer variables [19, 22]. Three-dimensional graphs of two-dimensional data are confusing and ought to be avoided.

The extra dimension is necessary, however, when presenting trivariate data. Wainer [24] stated that the third dimensions can cause ambiguity in graph interpretation and cautioned against using varying areas or volumes to depict additional variables. Stevens' Power Law predicts that lengths are unbiased, area judgments are biased, and volume judgments are even more biased [5]. Empirical studies have found that perceived area of a circle grows somewhat more slowly than the actual area: the reported perceived area = (actual area)^x, where x is around 0.8 [22]. Robbins [19] echoed this misperception of area and stated that the illusion is still more pronounced with volume. One recommendation for depicting more than two variables is using multiple display panels, with each panel showing bivariate relationships at discrete values of the third variable [19]. The multipanel trellis plot has been proposed as a viable candidate, but multipanel approaches can interject new problems when graphing three continuous variables (discussed below).

A number of options exist for plotting three-way relationships between variables, some more common in research practice than others. For purposes of exploratory or descriptive data analysis, the degree to which graphical methods provide a model-free representation of the data serves as a key desideratum for optimally useful graphs. We do not mean to suggest that a representation can exist entirely without conventions, assumptions, or underlying inferences. The important distinction is the extent to which the graph remains close to the raw data rather than relying on estimated numerical summaries to simplify and structure the graphical representation. The operating assumption is that during the exploratory stage of data analysis, a researcher does not yet understand the data well enough to decide on a model that adequately represents the data. As a result, optimally model-free graphs provide better choices for exploratory analysis early in the research process.

5.3 Graphical Options Ruled Out a Priori

This section briefly considers scatter plot matrices, line plots of conditional regression lines, and factorial-design style line plots. Each provides an excellent plot for some purposes, but not for model-free exploratory graphing of three continuous variables aimed at displaying the full three-way distribution.

Scatter Plot Matrix. The scatter plot matrix does a terrific job of showing bivariate relationships within a multidimensional set of data [5, 19]. A scatter plot matrix presents a matrix of bivariate scatter plots with individual variables defining the rows and columns. One can quickly scan a row or column and see how one variable relates to each of the other variables. Unfortunately, it cannot represent three-way distributions. The scatter plot matrix is akin to calculating a bivariate correlation matrix. Just as one needs to use another test to show two-way interactions when comparing more than two variables, one needs a different graph when ploting third variables. Figure 5.1 illustrates a scatter plot matrix using a simulated sample of 100 cases with three variables. The same data are plotted in Figures 5.2 to 5.6 for comparison.

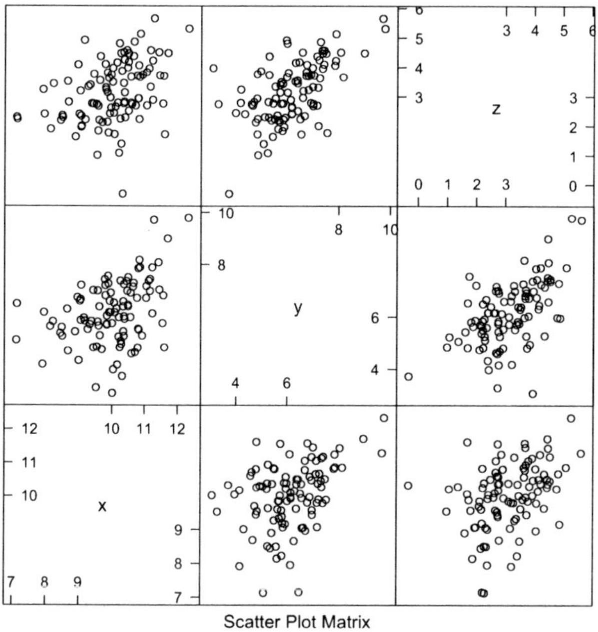

Fig. 5.1 Scatter plot matrix can only show bivariate relations within multivariate data.

Estimated Conditional Regression Lines. The multiple regression approach to showing interactions compares whether the slopes of estimated regression lines of the criterion variable on one predictor variable differ significantly when plotted at various critical values of the other predictor variable [1, 9] (see Figure 5.2). We did not use this approach, because the method is model dependent. The estimated regression lines plot the model, not the data, and thus do not indicate where along the regression lines the data fall. Moreover, the data must fit the model adequately or the conditional regression lines can present a misleading picture of the data.

One could lessen the model dependence by substituting nonparametric Lowess lines [4, 11] (also known as Loess lines), often used to provide a

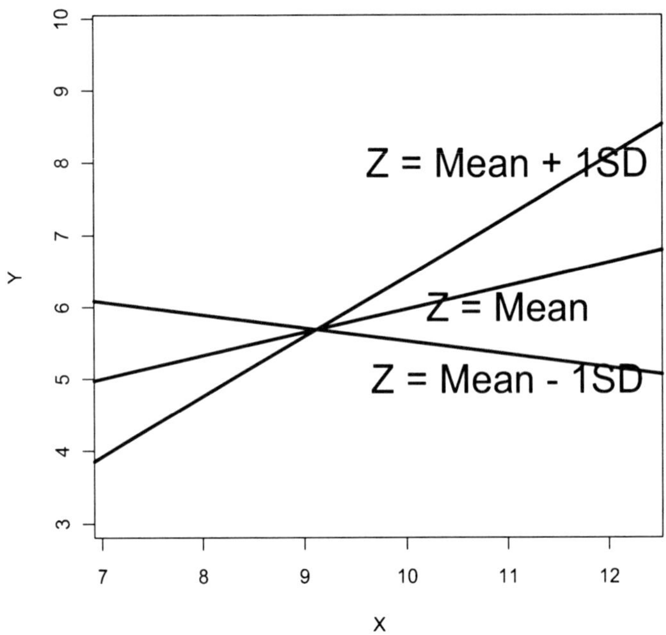

Fig. 5.2 Estimated conditional regression line approach to detecting interaction.

smoothed nonparametric regression to data showing a nonlinear relationship. Although Lowess lines do not assume the fit of a linear regression model, they still plot a nonparametric regression model rather than the data itself. Moreover, the result depends on the choice of smoothing parameters. For this reason, we ruled out this approach for present purposes.

Factorial Design Line Plots. A commonly used approach in studies with a factorial design graphs group means in a line plot (see Figure 5.3). Although frequently used to illustrate interactions, this approach would not be ideal, because it requires categorical data in the independent variable. Although one could divide continuous variables into groups or bins (commonly dichotomizing by the mean or median), thereby turning continuous variables into categorical variables, dichotomizing continuous variables is generally not recommended due to the resulting loss of information [8, 14, 1, 16].) Categorizing continuous variables results in less information loss than dichotomizing continuous variables but is also not recommended because it often makes the ANOVA model more complicated without having better fit [25]. The weakness of these model-based approaches to plotting is that they all require assumptions related to the model underlying the regression line or regression surface. If data do not satisfy the assumptions of the model, then the graphs resulting from the model can be misleading. In addition, they do not display variability around the line or surface without further modification.

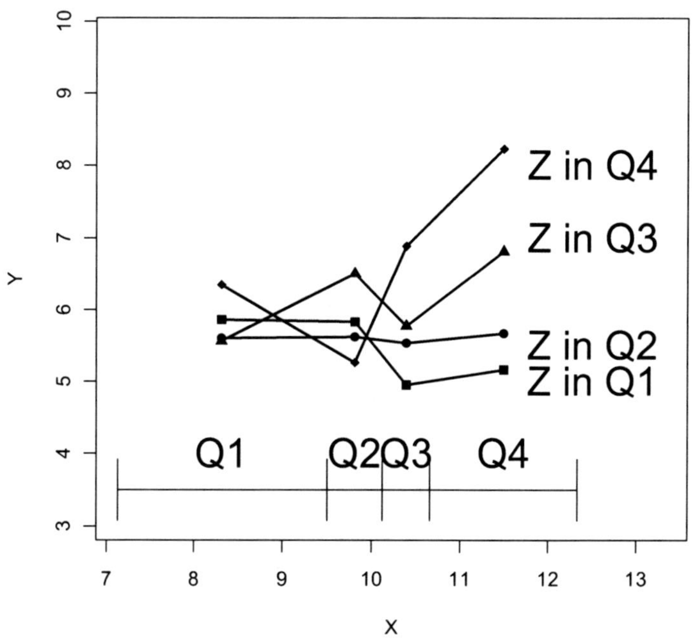

Fig. 5.3 Factorial design line plots approach to detecting interactions.

5.4 Plausible Graphical Alternatives

This section considers three-dimensional scatter plots, conditioning plots (coplots), and three-way bubble plots as relatively model-free means of graphing three continuous variables. A later section presents an empirical study evaluating these three methods.

3D Scatter Plot. Based on the aforementioned literature on plotting three-dimensional graphs on two-dimensional surfaces, one would expect users to have difficulty interpreting 3D scatter plots. Although the Moiré effect is ideally not present, the user is still being asked to make distinct judgments about distance and depth based on the illusion of an additional dimension. The 3D scatter plot shows main effects as increasing or decreasing plot heights across the axes and interaction as a different rate of height increase or decrease. Lines from the points to the graph floor help reduce ambiguity regarding the location of the points within the three-dimensional box.

Coplot. Cleveland introduced the coplot as a way to implement statistical control graphically without a statistical model, with the panels of a coplot presenting overlapping subsets of the data. This procedure works well for large data sets that can be subdivided without producing sparse subsets [10] (see Figure 5.5). The weakness associated with these graphs is similar to that of the line plot approach: The continuous variable represented by the

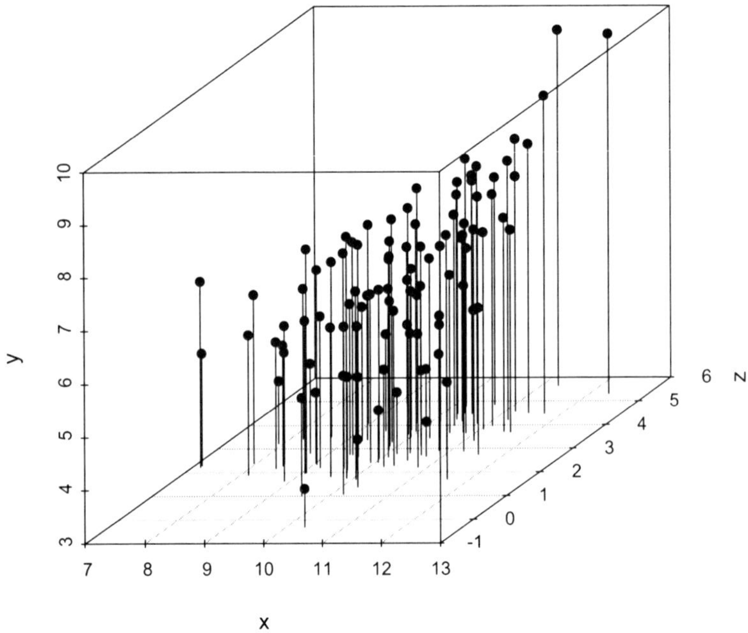

Fig. 5.4 3D scatter plot.

multiple panels is divided into overlapping categorical segments. One might need a large number of panels to adequately present the subdivided variable. Nonetheless, the literature describes the coplot as one of the better plots for presenting information with three variables [19, 10, 4]. The coplot shows main effects as positive or negative bivariate scatter plots across the panels and interaction as inconsistent bivariate relationships across the panels.

Three-Way Bubble Plot. Our motivation for creating the bubble plot extends from the need for a graph that can portray three continuous variables without resorting to inferring depth or relying on optical illusions. The concept of using a bubble plot this way is not new; examples can be found on web pages and in email list archives (including the R homepage). Nonetheless, we have been unable to locate any academic literature on this type of graph. In creating the bubble plot, we closely adhered to the guidelines proposed by Cleveland and Robbins. The resulting bubble plot is essentially an enhanced scatter plot, which, by itself, already conveys bivariate data clearly. Instead of rotating or tilting the axes to create the additional dimension, the size of the data points varies with the third variable. It is difficult to gauge point size differences unless they are lined up against an edge; "human beings are good at making comparisons with a straight line. Graphic comparisons are thus always easier when the quantities being compared start from a common base" [24, p. 33]. For this reason, the plot superimposes a muted grid to pro-

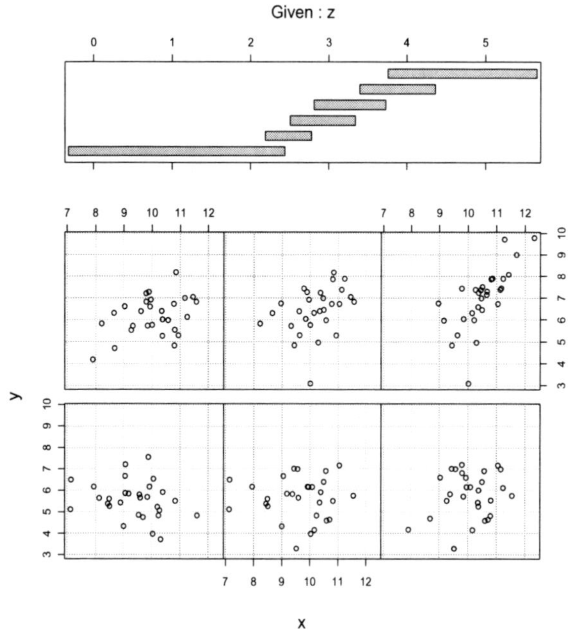

Fig. 5.5 Conditioning plot (coplot).

vide more reference lines for easier comparison. The bubble plot shows main effects as increasing or decreasing bubble sizes across the axes. Interactions effects appear as different rates of bubble size change for different values of one variable scanning across values of the other variable.

The bubble plot still contains some ambiguities. Using varying point size differences as the third variable bears a resemblance to using area comparisons, which humans do not perceive very accurately. In the bubble plot, the third variable may be perceived better because it is proportional to radius (and thus diameter) but not area. The problem of underestimating circle size differences is well documented. Although we added the grid to provide more references lines for easier comparisons of point sizes, the possibility that there is no valid comparison point along the same axis remains, which can make interactions hard to find. Data points that are close to one another often cluster or overlap, making the graph difficult to read. We used a conventional method of making the circles transparent to solve the problem of overlap. While using two-dimensional representations of three-dimensional spheres as graduated symbols is recommended in cartography, this method is impractical here due to the amount of overlap [21]. A further limitation is that while three-way bubble plots provide a useful display of the relative values of the third variable across the range of values for the other two variables, they are ineffective at conveying the absolute values on the third variable. While this

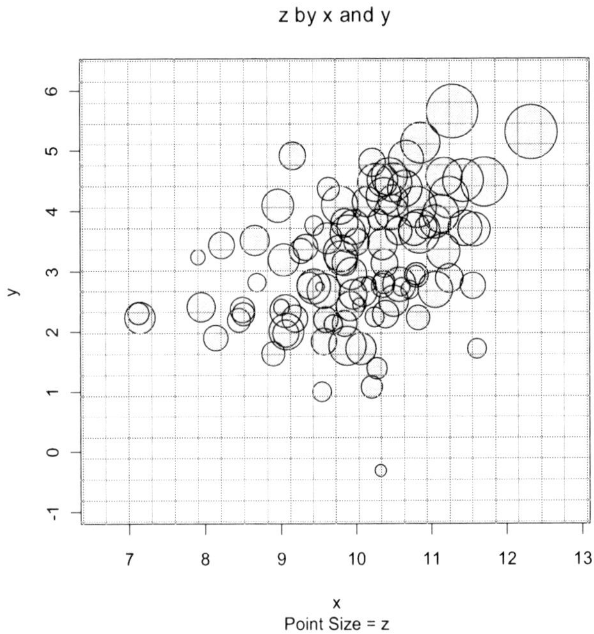

Fig. 5.6 Bubble plot.

is a limitation of the plot, it is also a reason that the usual cautions against relying on area perception do not directly apply to the three-way bubble plot. Judgments of absolute circle size do not play a role in reading the graph.

5.5 The *bp3way()* Function

> "If you are thinking about doing 'bubble plots' by `symbols(*, circles=*)`, you should really consider using `sunflowerplot` instead." (R core Development Team, *symbols()* function help file version 2.9.0)

The *bp3way()* function is not currently available as a contributed R package but rather as an R script. Running the script once makes the *pb3way()* function available for use. Saving a workspace after the function has been created makes it available for future use without rerunning the script. Help is currently included as documentation at the end of the script, but external to the function itself. An accompanying function, *bp.data()*, simulates data for use with the plot function. At this writing, the *bp3way()* function remains under development with version 3.2 as the current version.

As defined in the previous section, the *bp3way()* function plots graphs such that the values of two continuous variables are given by positions on the horizontal and vertical axes, and the value of a third variable is proportional to the radius of the plotted points. The function takes as an argument a matrix of data containing at least the variables to be plotted. Several additional parameters, discussed below, control various plot functions. As output, the function draws the plot in a graphics window and generates a list of plot-related values that can be used for further plotting or analysis. Optionally, plot parameters can also be output to the R console for review during an interactive R session. At the core of the function is the precise procedure warned against in the slightly too general advice quoted from an R help file at the head of this section. The last subsection of the previous section described the basic design and rationale for the three-way bubble plot. The remainder of this section describes the *bp3way()* function, outlines various options available within the *bp3way()* function, and briefly describes the accompanying *bp.data()* function.

5.5.1 *Use and Options of* bp3way() *Function*

The basic use of the *bp3way()* function closely parallels other graphical functions available as part of R. One must first place the variables one wants to plot into a data frame. The current version of *bp3way()* does not accept missing data. Cases with missing data cannot be plotted and should be removed from the data frame before using *bp3way()*. Simply calling the function will produce the graph. For example, the following line graphs the *trees* data set available in R — but it does not graph it very well.

```
>bp3way(trees)
```

However, the list output of the function can also be saved for further use through assignment to an R object.

```
>MyPlot <- bp3way(trees)
>MyPlot$rad.min
[1] 0.3075
```

5.5.2 *Six Key Parameters for Controlling the Graph*

The following six parameters are most central for obtaining the desired graph: *x, xc, bc, yc, names,* and *main*. The *x* parameter names the data frame, as in the above example where $x = \mathit{trees}$. The next three parameters make it convenient to permute the three variables one wishes to plot without changing the data frame: *xc, bc,* and *yc* give the column of the data frame containing the variable plotted on the horizontal axis, bubble radius, and vertical axis, respectively. The *names* parameter should contain the names of the variables in the order that they appear in the data frame and as they should appear in the graph. The graph resulting from the following call presents the data much better than did the default.

```
> bp3way(trees, xc = 1, bc = 3, yc = 2,
  names=c('Girth', 'Height', 'Volume'))
```

Finally, the *main* parameter allows for the addition of a title over the graph.

```
> bp3way(trees, xc = 1, bc = 3, yc = 2,
  names=c('Girth', 'Height', 'Volume'),
  main='Tree Volume by Girth and Height')
```

5.5.3 Additional Parameters Controlling the Data Plotted

In some cases, the plot may be easier to read if the x and y variables are standardized as standard normal scores. Setting *std* = TRUE will accomplish this. For large data sets, it may also work better to plot only a proportion of the data. This can be done by selecting cases at random, or by selecting from the beginning of the data set. For example, the following plots the first half of the *trees* data set, using a *z*-score scale. Note that standardization uses the entire data set and thus a sorted data set may produce all negative or all positive scores when only half are plotted.

```
> bp3way(trees, 1, 3, 2, names=c('Girth', 'Height',
  'Volume'), main='Tree Volume by Girth and Height',
  proportion = .5, random=FALSE, std=TRUE)
```

5.5.4 Parameters Controlling the Plotted Bubbles

A number of additional parameters allow further control of the precise appearance of the bubbles in the plot. The *x.margin* and *y.margin* control the space between the plot area and the edges of the horizontal and vertical axes, respectively. The *rad.ex* and *rad.min* parameters respectively control the size of the bubble radii proportional to the third variable and the minimum radius. Larger values of *rad.ex* make differences more pronounced, whereas a judiciously large value for *rad.min* prevents points with small values from disappearing from view due to their small size. Finally, the *fg* and *bg* parameters respectively control the color of the bubble edge and bubble fill.

5.5.5 Parameters Controlling the Grid

Nine parameters control the grid. The logical *grid* parameter turns the grid on (TRUE) or off (FALSE). The parameters *hlen* and *vlen* control the number of horizontal and vertical grid lines. The parameters *hlwd*, *vlwd*, *hlty*, *vlty*, *hcol*, and *vcol* control the line width, line type, and line color for the horizontal and vertical lines using values described in the help file for the *par()* function in R.

5.5.6 The tacit Parameter

The *tacit* parameter controls output to the R console. If set to FALSE, a call of the *bp3way()* function provides a list of key plot parameters to the console.

```
> bp3way(trees, 1,3,2, tacit=FALSE)
[1] Bubble Plot Parameters
[1]    Radius Expansion Factor: 1
[1]    Minimum Radius: 0.3075
[1]    X Margin: 0.1
[1]    Y Margin: 0.1
[1]    Plotted Proportion: 1
[1]    Standardized: FALSE
```

These six parameters are data sensitive if left to their default values. As such, the *tacit* parameter makes it more convenient to monitor these when graphing data. The values can also be used to choose user-specified values to override the default values. This is purely a convenience feature intended for tweaking unsatisfactory plots and defaults to TRUE.

5.5.7 The bp.data() Function

The *bp.data()* function is a very simple function that generates a data frame containing three variables that can be used with *bp3way()*. The underlying model regresses the third variable on the first two. User-specified parameters control the sample size (N), mean and standard deviation of the first predictor (MX and SDX), mean and standard deviation of the second predictor (MY and SDY), error variance added to the outcome variable ($Berror$), regression weights for the intercept, X, Y, and the interaction term (a, b, c, d), and the amount of shared variance between X and Y (SV). The following call illustrates both the use of the function and the default values.

```
MyData <- bp.data(N=5000, MX=10, SDX=2, MY=10, SDY=2,
                  Berror=6, a=10, b=1, c=.5, d=.5, SV=1)
```

The present section has outlined the use of the *bp3way()* function to construct three-way bubble plots in R. The next section describes an empirical study exploring the relative success of three-way bubble plots, 3D scatter plots, and coplots as a means of representing relationships between three continuous variables.

5.6 An Empirical Study of Three Graphical Methods

The design features and default values of the *pb3way()* function described above reflect an exploratory trial-and-error process of plotting data as three-way bubble plots with the researchers' judgments supplying the main source of feedback and evaluation. The empirical study was designed with two goals: (1) to gather empirical data to assess the approach for communicating three-way relationships between continuous variables and (2) to gather empirical data to evaluate this approach in comparison to existing alternatives. Based on the literature, we hypothesized that both three-way bubble plots and coplots would outperform 3D scatter plots.

5.6.1 Method

This section describes the participants, design, materials, and procedure.

Participants. One hundred and eight undergraduate students taking Psychology 101 at John Jay College of Criminal Justice participated in the study during the Fall 2008 semester. Most of the students had not taken statistics courses, so their experiences in interpreting graphs were probably similar to those of an average person.

Design. The study compared three graph types (three-way bubble plot, 3D scatter plot, and coplot) across six data sets. The study used a mixed within- and between-subject design, with each participant seeing only one of three graph types (between) but seeing each of six data conditions with various degrees of main effects and interactions (within). Students were randomly assigned to the three graph conditions. The six data conditions had variations of positive and negative main effects and interactions (Appendix A).

Materials. We created all three types of plots using R software and printed them on 8.5-by-11-inch paper (see Appendixes B to D). Because most participants had little experience with statistics and little experience reading the graphs in question, we labeled the axes with concrete and readily understood variable names. We renamed the dependent variable sales and the independent variables as staff size and number of stores, respectively. The selected concrete variables did not relate to each other in obvious meaningful relationships, such as sales and price or sales and location.

The questions assessed how staff size and sales related to one another (main effect), whether one main effect was stronger than another (relationship between staff size and sales versus relationship between store numbers and sales), whether the relationship between staff size depends on store numbers (interaction), the confidence ratings of each answer, the clarity rating of the graph, and the graph's ease-of-use rating. The first question asked "Is the relationship between staff size and sales positive or negative? Circle corresponding option below" (*Positive, Negative, Unsure*). A follow-up question asked "How confident are you of the answer?" (5-point Likert scale from *Not Confident* to *Very Confident*). The second question asked "Is the relationship between staff size and sales stronger or weaker than the relationship between number of stores and sales? Circle corresponding option below" (*Stronger, Weaker, Unsure*) and was followed by an identical confidence question. Question three asked "Does the relationship between staff size and sales depend upon number of stores? Circle corresponding option below" (*Yes, No*), with the same confidence question. The fourth and fifth questions asked "Rate the clarity, by which the graph conveyed the information" (from *Not Clear* to *Very Clear*) and "Rate the graph's ease of use" (from *Very Difficult* to *Very Easy*) each using a 5-point Likert scale. There was no restriction on the amount of time participants had for reading the plots and answering the questions but all participants completed the study within 60 minutes.

Procedure. Participants completed the study in small groups in an on-campus psychology laboratory. Each participant signed an informed consent form first, then interpreted a packet of six plots and answered several questions regarding each graph, and finally received an educational debriefing form.

We examined the effects of plot type and data type on four dependent variables: accuracy, confidence, clarity ratings, and ease-of-use ratings. Accuracy represented the proportion of questions answered correctly for each graph. Omitted answers were marked as incorrect, but these accounted for less than 2 percent of the data, with no more than 2 missing responses for any one question (N = 108). For the multivariate analysis, this was rescaled to range from 0 to 3 as the number of correct answers rather than a proportion. Confidence corresponded to the mean of each participant's confidence rating of his or her answers averaged across the questions and thus ranged from 1 to 5. Clarity and ease simply reflect the 1 to 5 Likert scale responses. The study included the latter two measures based on the rationale that clarity constitutes a necessary but insufficient condition for ease of use, and thus the two constructs differ conceptually however much they may correlate empirically.

5.6.2 Results

Although we used R software to create the graphs and data conditions, we used Mplus v.5.21 for the multilevel linear mixed-effects analysis because there is no straightforward way to run the analysis with ordinal level outcomes in R. Descriptive statistics provided a check for assumptions of univariate normality, kurtosis, and the minimum observations requirement for each cell. Box's Test of Equality of Covariance Matrices provided an assessment of the homogeneity of variance across the cells. Mahalanobis Distance provided a check for multivariate outliers. Pearson correlation matrices provided a check for multicollinearity.

The multilevel mixed-effects models analyzed the data in two stages. The overall multivariate analysis compared performances between the three graph types (bubble plot, 3D scatter plot, coplot) and within the six data conditions (positive main effects with positive interaction, positive and negative main effects with no interaction, no main effect with positive interaction, no main effect with no interaction, positive main effects with negative interaction, no main effect with negative interaction) using the Robust Maximum-Likelihood estimator [17]. The six data conditions were nested within the 108 participants for a total of 648 observations. Graph type and data conditions were recoded using effects coding for conducting the multilevel mixed-effects analysis, using the first plot type (bubble plot) and study condition (main effects with positive interaction) as the reference groups. Since the multivariate analysis showed statistical significance, protected follow-up univariate Exact Wilcoxon Mann–Whitney rank sum tests checked for simple effects between graph conditions within each data condition for the three ordinal variables and protected t-tests checked for simple effects between graphs for the continuous variable.

Descriptive statistics of the outcome variables (accuracy, confidence, clarity, and ease-of-use) revealed 4 missing cases in the accuracy condition for the 108 participants, but the number of missing data points was less than 5% of the total data and the omitted answers were treated as incorrect for that variable. The three levels of the independent variable yielded 36 observations in each cell across the dependent variables, which satisfied the minimum requirements of 20 observations for each cell. Most of the cell distributions had minimal skew (within two SEs of zero; see Figure 5.7). The 3D scatter plot condition had accuracy and ease-of-use outcomes with skewness near 2.50 SEs from zero. Kurtosis was within two SEs of zero and not problematic for any of the cell distributions. Evaluation of Mahalanobis Distances for each case against $\chi^2(1, N = 108) = 10.828$ found no multivariate outliers (all $p > .001$). Averaged across the six data conditions, the participants interpreted the coplot ($M = .53$, $Mdn = .53$, $SD = .15$) most accurately and the 3D scatter plot ($M = .43$, $Mdn = .44$, $SD = .11$) least accurately (see Table 5.1). Participants' confidences in their answers for coplot ($M = 3.56$, $Mdn = 3.64$, $SD = .59$) were the highest and 3D scatter plot ($M = 3.49$,

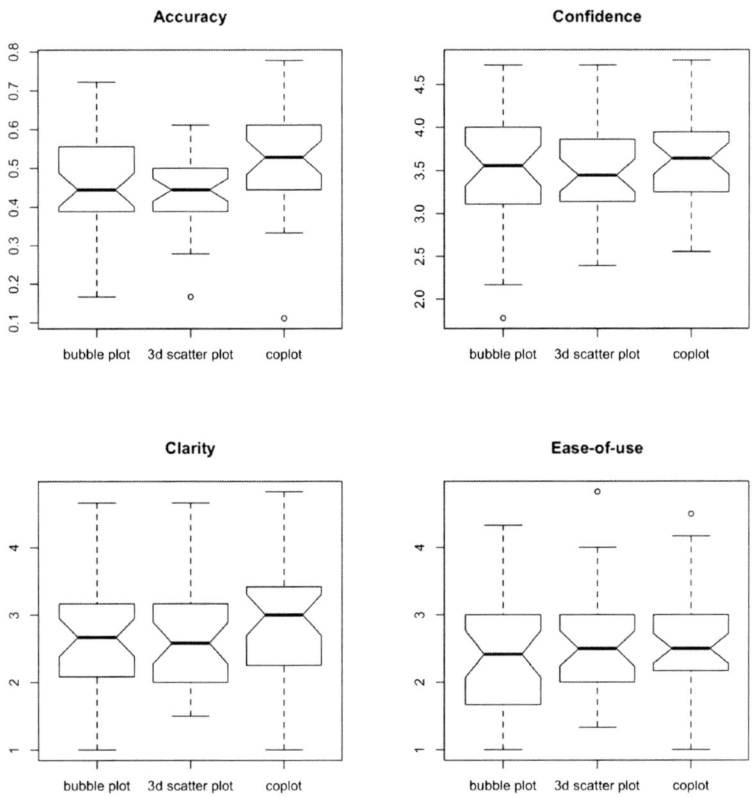

Fig. 5.7 Clustered boxplots of plot conditions within each outcome.

Mdn = 3.44, *SD* = .53) the lowest, but they rated all graph types as between neither-confident-nor-unconfident and confident (see Table 5.2). Coplot (*M* = 2.84, *Mdn* = 3.00, *SD* = .81) appeared the clearest and bubble plot (*M* = 2.64, *Mdn* = 2.67, *SD* = .94) the least clear, but all three graphs were rated between not clear to neither-clear-nor-unclear on average (see Table 5.3). The participants also found coplot (*M* = 2.62, *Mdn* = 2.50, *SD* = .77) easiest to use and bubble plot (*M* = .2.39, *Mdn* = 2.41, *SD* = .98) most difficult to use, but all three graphs were between difficult and neither-easy-nor-hard to use (see Table 5.4). There were no floor or ceiling effects, as the outcome variables appeared to contain ample variability. Pearson correlation matrices at the individual and group level indicated that the outcome variables were not sufficiently correlated to suggest multicollinearity (see Table 5.5).

Of the four outcome variables, accuracy, clarity, and ease-of-use were treated as ordinal variables, whereas confidence was treated as a continuous

Table 5.1 Mean, median, and standard deviation of accuracy by graph type and data set

	Overall					Data Set					
Condition	*n*	*M*	*Mdn*	*SD*	*Skew*	*1*	*2*	*3*	*4*	*5*	*6*
Bubble	33	0.46	0.44	0.12	-0.14	0.68[a]	0.43	0.41	0.49[c]	0.33[ef]	0.42[g]
3D Scatter	35	0.43	0.44	0.11	-2.38	0.58	0.40[b]	0.39	0.40[d]	0.56[e]	0.28[g]
Coplot	36	0.53	0.53	0.15	-0.64	0.55[a]	0.51[b]	0.48	0.71[cd]	0.55[f]	0.40

Exact Wilcoxon Mann–Whitney rank sum tests: a: $z = 2.08$, $p = .038$; b: $z = -1.99$, $p = .046$; c: $z = -2.53$, $p = .011$; d: $z = -3.37$, $p < .001$; e: $z = -3.24$, $p = .001$; f: $z = -2.79$, $p = .004$; g: $z = 2.05$, $p = .042$
Skew = skewness z-score

Table 5.2 Mean, median, and standard deviation of confidence by graph type and data set

	Overall					Data Set					
Condition	*n*	*M*	*Mdn*	*SD*	*Skew*	*1*	*2*	*3*	*4*	*5*	*6*
Bubble	36	3.49	3.56	0.65	-1.24	3.71	3.37[a]	3.58	3.36	3.37	3.57
3D Scatter	36	3.49	3.44	0.53	0.35	3.73	3.56	3.54	3.34	3.39	3.41
Coplot	36	3.56	3.64	0.59	-0.31	3.59	3.72[a]	3.44	3.48	3.59	3.56

Two-tailed t tests with equal variance assumed: a: $t(70) = -2.07$, $p = .042$. In two instances where equal variances were questionable, a Welch t test was performed, neither rejected the null hypothesis.
Skew = skewness z-score

Table 5.3 Mean, median, and standard deviation of clarity by graph type and data set

	Overall					Data Set					
Condition	*n*	*M*	*Mdn*	*SD*	*Skew*	*1*	*2*	*3*	*4*	*5*	*6*
Bubble	36	2.64	2.67	0.94	-0.03	2.56	2.78	2.78	2.42	2.61	2.72
3D Scatter	36	2.68	2.58	0.78	1.62	2.58	2.53[a]	2.75	2.69	2.69	2.83
Coplot	36	2.84	3.00	0.81	0.09	2.53	3.25[a]	2.61	2.58	3.17	2.92

Exact Wilcoxon Mann–Whitney rank sum tests: a: $z = -2.68$, $p = .007$
Skew = skewness z-score

Table 5.4 Mean, median, and standard deviation of ease-of-use by graph type and data set

	Overall					Data Set					
Condition	*n*	*M*	*Mdn*	*SD*	*Skew*	*1*	*2*	*3*	*4*	*5*	*6*
Bubble	36	2.39	2.42	0.98	0.49	2.08	2.33	2.50	2.22	2.61	2.61
3D Scatter	36	2.52	2.50	0.74	2.49	2.36	2.36[a]	2.61	2.50	2.64	2.67
Coplot	36	2.62	2.50	0.77	0.68	2.31	2.86[a]	2.50	2.36	3.11	2.58

Exact Wilcoxon Mann–Whitney rank sum tests: a: $z = -2.11$, $p = .035$
Skew = skewness z-score

Table 5.5 Intercorrelation between outcome measures

Variables	1	2	3	4
1. Accuracy	1	.061 [-.016, .137]	-.058 [-.134, .019]	−.102** [-.177, -.025]
2. Confidence	.106 [-.089, .292]	1	.497** [.418, .536]	.471** [.408, .529]
3. Clarity	-.118 [-.304, .076]	.586** [.446, .698]	1	.784** [.752, .812]
4. Ease-of-use	-.115 [-.301, .080]	.562** [.417, .679]	.866** [.810, .907]	1

Note. Pearson correlations at the observation level (N = 648) above the diagonal; Pearson correlations of the mean across data set for individual participants (N = 108) below the diagonal. 95 percent confidence interval included in the brackets below the correlation.
** Correlation is significant at the 0.01 level (2-tailed).

variable. Accuracy was the proportion of questions participants interpreted correctly. Because there were three questions per graph, the participants could only have four possible accuracy outcomes: 0, .33, .66, 1.00, so accuracy was considered ordinal for the analyses. Clarity and ease-of-use were 1 to 5 Likert ratings. Although researchers often consider summative response scales, like the Likert scale, as between ordinal and interval scales, we treated both variables as ordinal because each scale only had five options, and each variable consisted of only one scale. Confidence was treated as a continuous variable because it averaged across three 5-point Likert scales, producing 13 possible values ranging between 1 and 5.

Each outcome was analyzed separately. The mixed-level models included random intercept but no random slopes. The models included both main effects and interactions.

$$\text{Level 1: } \mathbf{O}^* = \alpha_0 + \alpha_1{}'\mathbf{D} + \alpha_2'\mathbf{DG} + \mathbf{u}_1$$
$$\text{Level 2: } \alpha_0 = \beta_0 + \beta_1'\mathbf{G} + \mathbf{u}_2$$
$$\text{Threshold Model:} O = \{0 \text{if } \mathbf{O}^* \leq \tau_1,\ 1 \text{ if } \tau_1 < \mathbf{O}^* \leq \tau_2, \ldots\ k \text{ if } \tau_{k-1} < \mathbf{O}^* \leq \tau_k\}$$

O is the observed ordered-categorical outcome score, $\mathbf{O}^*$ is a vector of latent continuous outcome scores for each observation, $\mathbf{D}$ is a matrix of effect-coded dichotomous variables representing the data set condition, $\mathbf{G}$ is a matrix of dummy-coded dichotomous variables representing the graph condition, $\mathbf{DG}$ is a matrix of data-by-graph interaction terms, α and β represent vectors of linear weights, the $\mathbf{u}$ variables represents vectors of residuals, and the τ_s represent fixed thresholds.

There was a statistically significant main effect of coplot being more accurate than the overall mean for accuracy (see Table 5.6). The statistically significant main effect for condition 5 suggested that participants on average interpreted data condition for positive main effects with negative interaction less accurately than overall mean accuracy. There were a few statistically significant graph-by-data interactions. Participants interpreted 3D scatter plots showing positive main effects with negative interaction (condition 5) more accurately than overall mean accuracy and no main effect with negative interaction (condition 6) less accurately. The participants interpreted coplots showing no main effects with interactions (condition 4) more accurately than overall mean accuracy, and positive main effects with negative interaction (condition 5) more accurately.

Table 5.6 Multilevel mixed-effects model for accuracy

MODEL RESULTS	Estimate	S.E.	Two-Tailed p-Value
Within Level			
D2	-0.134	0.295	0.651
D3	-0.275	0.161	0.087
D4	0.156	0.363	0.666
D5	-0.850	0.295	0.004*
D6	-0.247	0.268	0.356
D2G2	-0.029	0.399	0.941
D2G3	-0.001	0.379	0.997
D3G2	-0.039	0.320	0.904
D3G3	-0.077	0.306	0.802
D4G2	-0.498	0.541	0.358
D4G3	1.164	0.519	0.025*
D5G2	1.744	0.426	0.000*
D5G3	0.923	0.437	0.035*
D6G2	-0.767	0.388	0.048*
D6G3	-0.649	0.405	0.109
Between Level			
G2	-0.196	0.184	0.286
G3	0.520	0.209	0.013*
Thresholds			
accuracy$1	-1.715	0.159	0.000
accuracy$2	0.359	0.131	0.006
accuracy$3	2.057	0.177	0.000
Residual Variance			
accuracy	0.100	0.115	0.388

Note: D = effect-coded dichotomous variables representing the data set condition; G = dummy-coded dichotomous variables representing the graph condition; DG = data-by-graph interaction terms.

Because the multivariate analysis for accuracy was statistically significant, univariate Exact Wilcoxon Mann–Whitney rank sum tests were performed to check for mean differences between graph types in each data condition. The participants interpreted the bubble plot more accurately than coplot for positive main effects with positive interaction ($z = 2.08$, $p = .038$); 3D scatter plot less accurately than coplot for positive and negative main effects with no interactions ($z = -1.98$, $p = .046$); both bubble plot and 3D scatter plot less accurately than coplot for no main effect with no interaction ($z = -2.53$, $p = .011$; $z = -3.37$, $p < .001$); bubble plot less accurately than 3D scatter plot ($z = -3.24$, $p = .001$) and coplot ($z = -2.79$, $p = .004$) for positive main effects with negative interaction; bubble plot more accurately than 3D scatter plot ($z = 2.05$, $p = .042$; see Figure 5.8).

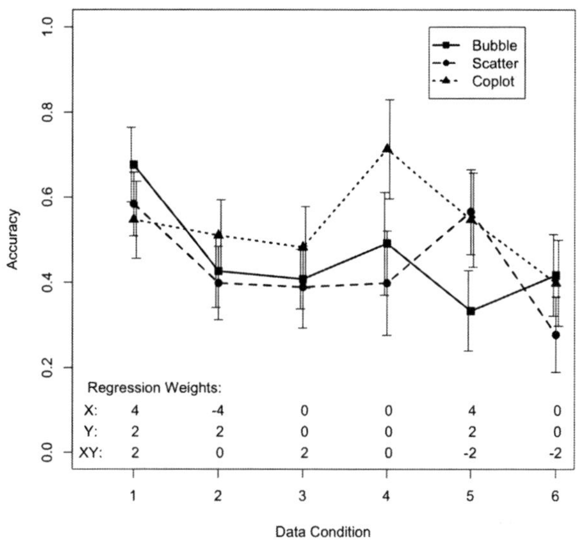

Fig. 5.8 Mean accuracy and standard error by graph type and data set.

The confidence outcome, measuring participants' confidence in their interpretations, also showed no statistically significant difference between the graphing conditions. There was no statistically significant main effect, but the coplot by condition 3 interaction was statistically significant, suggesting participants were more confident than the overall mean confidence in interpreting coplot for no main effects with positive interaction and less confident in interpreting coplot for no main effect with positive interaction (see Table 5.7). The residual variance of 0.34 was low compared to the standard deviations, suggesting the model accounted for a good portion of the variation in effects across participants. The correlations between the effects were generally low, suggesting no strong dependencies between parameters. There were good correspondences between the parameters and the p-values, suggesting there were no serious problems with lack of power for certain effects.

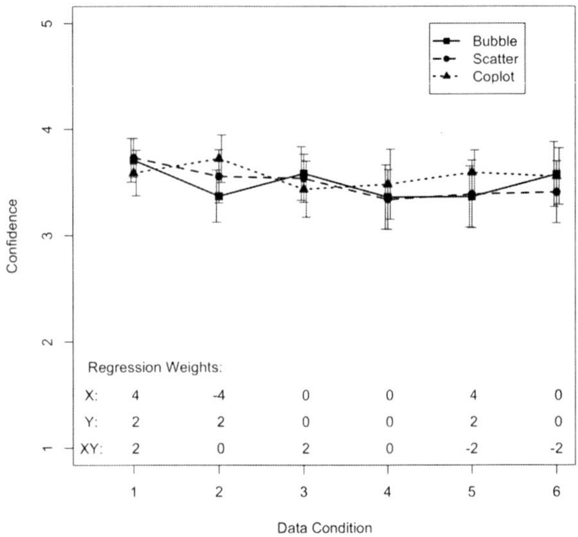

Fig. 5.9 Mean confidence and standard error by graph type and data set.

Confidence was scored on a Likert scale of 1 to 5 (higher rating indicating higher confidence), thus the intercept of 3.49 suggested the grand mean for confidence was between neither-confident-nor-not-confident and confident. The standard errors were between .07 and .14, which made sense in relation to the size of the effects. Because the confidence variable was treated as a continuous variable, univariate t-tests were performed to check for mean differences between graph types in each data condition. Participants were less confident in their interpretations of bubble plot than coplot ($t(70) = -2.07$, $p = .042$) for positive and negative main effects with no interaction (condition 2, see Figure 5.9). In two instances where equal variances were questionable, Welch t-tests were performed, but neither rejected the null hypothesis.

The clarity outcome also did not have statistically significant differences between the graphing conditions and overall mean clarity. Participants found data condition showing no main effect with no interaction clearer than mean clarity across conditions (see Table 5.8). Several interactions showed statistical significance. Participants found 3D scatter plots depicting positive and negative main effects with no interaction (condition 2) less clear. They read coplots showing no main effects with positive interaction (condition 3) more accurately and positive main effects with negative interactions (condition 5) more accurately. Univariate Exact Wilcoxon Mann–Whitney rank sum tests were performed to check for mean differences between graph types in each

Table 5.7 Multilevel mixed-effects model for confidence

MODEL RESULTS	Estimate	S.E.	Two-Tailed *p*-Value
Within Level			
D2	-0.123	0.085	0.148
D3	0.090	0.086	0.300
D4	-0.133	0.073	0.070
D5	-0.128	0.071	0.073
D6	0.080	0.091	0.378
D2G2	0.186	0.137	0.175
D2G3	0.283	0.127	0.026*
D3G2	-0.046	0.134	0.733
D3G3	-0.217	0.113	0.054*
D4G2	-0.022	0.115	0.846
D4G3	0.052	0.120	0.666
D5G2	0.024	0.127	0.850
D5G3	0.158	0.097	0.103
D6G2	-0.166	0.133	0.212
D6G3	-0.087	0.134	0.515
Residual Variance			
confidence	0.342	0.030	0.000
Between Level			
G2	-0.001	0.138	0.996
G3	0.069	0.144	0.634
Intercepts			
confidence	3.494	0.107	0.000
Residual Variance			
confidence	0.284	0.045	0.000

Note: D = effect-coded dichotomous variables representing the data set condition; G = dummy-coded dichotomous variables representing the graph condition; DG = data-by-graph interaction terms.

data condition. Participants found 3D scatter plots less clear than coplots in condition 2 ($z = -2.68$, $p = .007$; see Figure 5.10).

The ease-of-use outcome did not show statistically significant differences between the graphing conditions and the overall mean ease-of-use. There was a statistically significant main effect at condition 6, meaning participants found no mean effect and negative interaction easier to use than overall mean ease-of-use (see Table 5.9). There was no statistically significant graph type by data interaction. Univariate Exact Wilcoxon Mann–Whitney rank sum tests were performed to check for mean differences between graph types in each data condition. In condition 2, participants found coplots easier to use than bubble plots ($z = -2.11$, $p = .035$) and 3D scatter plots ($z = -2.11$, $p = .036$; see Figure 5.11).

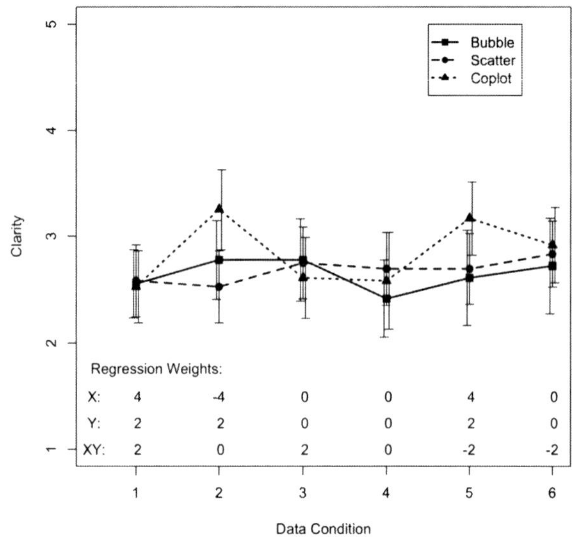

Fig. 5.10 Mean clarity and standard error by graph type and data set.

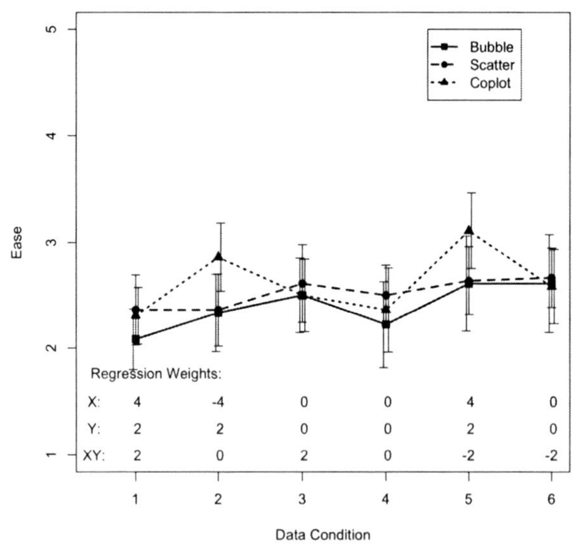

Fig. 5.11 Mean ease-of-use and standard error by graph type and data set.

Table 5.8 Multilevel mixed-effects model for clarity

MODEL RESULTS	Estimate	S.E.	Two-Tailed *p*-Value
Within Level			
D2	0.401	0.287	0.162
D3	0.447	0.266	0.094
D4	-0.622	0.303	0.040*
D5	-0.099	0.310	0.749
D6	0.241	0.329	0.464
D2G2	-0.751	0.366	0.040*
D2G3	0.631	0.435	0.147
D3G2	-0.263	0.351	0.453
D3G3	-0.928	0.392	0.018*
D4G2	0.669	0.408	0.101
D4G3	-0.193	0.531	0.717
D5G2	0.150	0.400	0.708
D5G3	0.897	0.430	0.037*
D6G2	0.102	0.407	0.803
D6G3	-0.028	0.427	0.948
Between Level			
G2	0.166	0.535	0.756
G3	0.489	0.559	0.382
Thresholds			
clarity$1	-2.533	0.479	0.000
clarity$2	-0.156	0.441	0.724
clarity$3	1.991	0.467	0.000
clarity$4	4.430	0.532	0.000
Residual Variance			
clarity	4.154	0.969	0.000

Note: D = effect-coded dichotomous variables representing the data set condition; G = dummy-coded dichotomous variables representing the graph condition; DG = data-by-graph interaction terms.

5.7 Discussion

Other than accuracy, for which graph type had a statistically significant main effect with coplots being more accurately interpreted than the other plots, most of the differences occurred in various graph-by-data interactions for the other outcomes. Although preliminary, this finding hints that the optimal plot is data dependent. Depending on whether researchers want to show data with main effects or interactions, one type of graph may be better than another. Data conditions 1, 3, 5, and 6 had graphs containing interactions, and the bubble plot performed comparably to the coplot in each of these conditions with two exceptions. For accuracy of positive main effects with a negative interaction, the bubble plot performed worse than the other two plots. When all three effects were positive, the bubble plot was read more

Table 5.9 Multilevel mixed-effects model for ease-of-use

MODEL RESULTS	Estimate	S.E.	Two-Tailed p-Value
Within Level			
D2	-0.161	0.337	0.633
D3	0.359	0.284	0.206
D4	-0.542	0.353	0.125
D5	0.571	0.358	0.110
D6	0.646	0.299	0.031*
D2G2	-0.251	0.448	0.575
D2G3	0.778	0.415	0.060
D3G2	-0.128	0.407	0.753
D3G3	-0.669	0.378	0.077
D4G2	0.537	0.423	0.205
D4G3	-0.314	0.531	0.555
D5G2	-0.268	0.457	0.557
D5G3	0.669	0.480	0.164
D6G2	-0.271	0.399	0.498
D6G3	-0.635	0.380	0.094
Between Level			
G2	0.613	0.591	0.300
G3	0.804	0.602	0.182
Thresholds			
easeuse$1	-1.835	0.532	0.001
easeuse$2	0.647	0.504	0.199
easeuse$3	3.087	0.537	0.000
easeuse$4	5.212	0.577	0.000
Residual Variance			
easeuse	4.744	1.061	0.000

Note: D = effect-coded dichotomous variables representing the data set condition; G = dummy-coded dichotomous variables representing the graph condition; DG = data-by-graph interaction terms.

accurately than the coplot. The research sought to evaluate the bubble plot against other model-free plots for showing interactions and the usefulness of the bubble plot appears to fall between the 3D scatter plot and the coplot. Of the three alternatives, the coplot emerged as the slight favorite. Participants interpreted coplots more accurately than the other two plots, despite not feeling more confident in their answers, finding it clearer or easier to use. In addition, the coplot was found to be the best or next to best alternative in almost all of the univariate comparisons of graph types within conditions.

Based on the results of this study, our recommendation is to use at least two types of graphs for exploratory analysis, especially because one cannot predict ahead of time which graph would work best. It appears useful to view data in different ways, even if one graph works consistently better than the others. If a model fits the data, one should report results using model-based graphs. At present, model-free graphs probably have the most use during the exploratory stage of analysis.

The bubble plot certainly was not inferior to other model-free plots for showing interactions. It appears to show positive interactions somewhat more accurately than negative interactions. Future bubble plot studies should examine other variations of main effects and interactions to see whether these findings generalize. Aside from accuracy, the other three dependent variables showed little variability. Thus, future studies should seek to improve the sensitivity of the outcome measures relevant to graphing. Much remains to be learned about the cognition and perception of these three graph types. More detailed perceptual theory is needed to optimize graph design. Perhaps it is time to redirect focus back to understanding how the human visual system can better perceive contemporary three-dimensional graphs, especially because computers and software packages can now create three-dimensional graphs much more easily than before. Identifying additional factors that affect graph design would allow us to modify the design of current model-free graphs to maximize the effectiveness for all three graph types.

Acknowledgements We would like to thank Frances Figueroa for assistance in collecting the data reported in this chapter.

Appendix A

Study Conditions

Graph 1: bubble plot

Graph 2: 3D scatter plot

Graph 3: coplot

Data 1: positive main effects, positive interaction $Y = 10 + 4X + 2Z + 2XZ + \text{error}$

Data 2: positive and negative main effects, no interaction $Y = 10 - 4X + 2Z + \text{error}$

Data 3: no main effect, positive interaction $Y = 10 + 2XY + \text{error}$

Data 4: no main effect, no interaction $Y = 10 + \text{error}$

Data 5: positive main effects, negative interaction $Y = 10 + 4X + 2Z - 2XZ + \text{error}$

Data 6: no main effect, negative interaction $Y = 10 - 2XY + \text{error}$

Appendixes B and C

Bubble plot conditions 1 to 6

3D scatter plot conditions 1 to 6

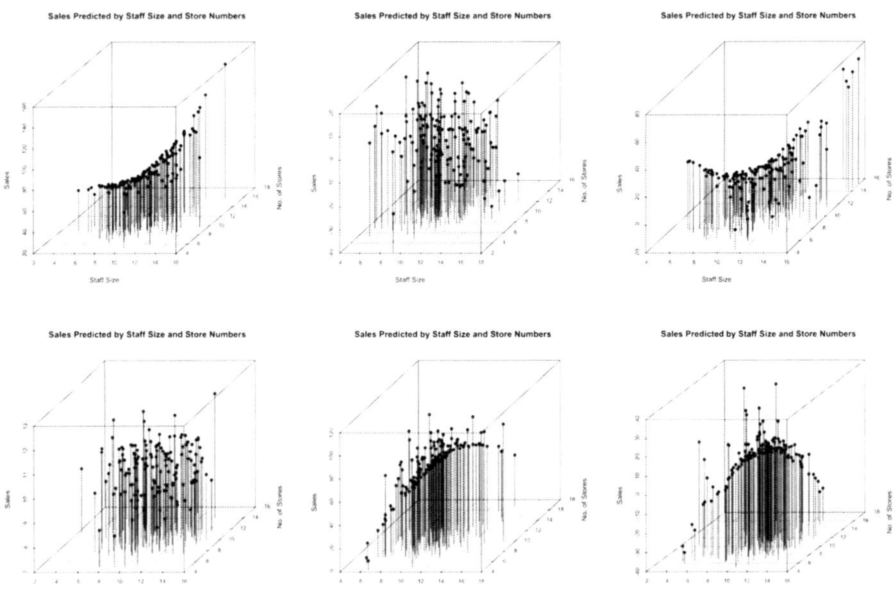

Appendix D

Coplot conditions 1 to 6

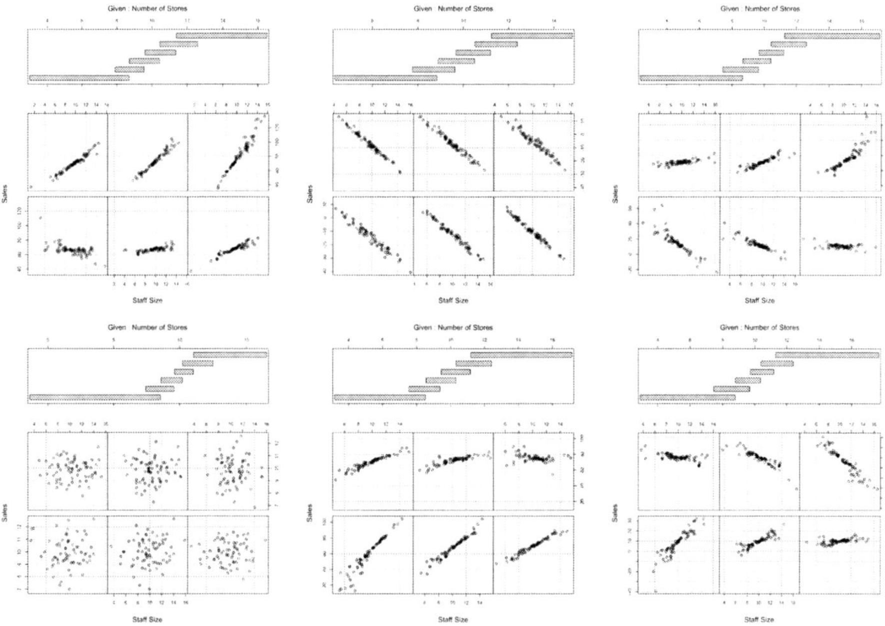

References

1. Aiken, L.S., West, S.G.: Multiple Regression: Testing and Interpreting Interactions. Sage Publications, Inc, Newbury Park (1991)
2. Baron, R.B., Kenny, D.A.: The moderator-mediator variable distinction in social psychological research: conceptual, strategic, and statistical considerations. Journal of Personality and Social Psychology **51**(6), 1173–1182 (1986)
3. Cleveland, W.S.: The Elements of Graphing Data. Wadsworth Advanced Books and Software, Monterey (1985)
4. Cleveland, W.S.: Visualizing Data. Hobart press, Summit, New Jersey (1993)
5. Cleveland, W.S.: The Elements of Graphing Data, revised edn. Hobart Press, Summit, NJ (1994)
6. Cleveland, W.S., Harris, C.S., McGill, R.: Experiments on quantitative judgments of graphics and maps. The Bell System Technical Journal **62**(6), 1659–1674 (1982)
7. Cleveland, W.S., McGill, R.: An experiment in graphical perception. International Journal of Man-Machines Studies **25**(5), 491–500 (1986)
8. Cohen, J.: The cost of dichotomization. Applied Psychological Measurement **7**(3), 249–253 (1983)
9. Cohen, J., Cohen, P., West, S., Aiken, L.: Applied multiple regression/correlation analyses for the behavioral sciences, 3rd edn. Lawrence Erlbaum, Hillsdale, NJ (2002)
10. Fox, J.: An R and S-Plus Companion to Applied Regression. Sage Publications, Thousand Oaks (2002)

11. Fox, J.: car: Companion to Applied Regression. R package version 1.2-14 (2009). URL `http://CRAN.R-project.org/package=car`. I am grateful to Douglas Bates, David Firth, Michael Friendly, Gregor Gorjanc, Spencer Graves, Richard Heiberger, Georges Monette, Henric Nilsson, Derek Ogle, Brian Ripley, Sanford Weisberg, and Achim Zeileis for various suggestions and contributions
12. Hartwig, F., Dearing, B.E.: Exploratory data analysis. Sage university paper series on quantitative applications in the social sciences, series no. 07-016. Sage Publications, Beverly Hills (1979)
13. Kraemer, H.C., Kiernan, M., Essex, M., Kupfer, D.J.: How and why criteria defining moderators and mediators differ between baron & kenny and macarthur approaches. Health Psychology **27**(2), S101–S108 (2008)
14. MacCallum, R.C., Zhang, S., Preacher, K.J., Rucker, D.D.: On the practice of dichotomizing of quantitative variables. Psychological Methods **7**, 19–40 (2002)
15. MacKinnon, D.P., Fairchild, A.J., Fritz, M.S.: Mediation analysis. Annual Review of Psychology **58**, 593–614 (2007)
16. Maxwell, S.E., Delaney, H.D.: Bivariate median splits and spurious statistical significance. Psychological bulletin **113**(1), 181–190 (1993)
17. Muthen, L.K., Muthen, B.O.: Mplus user's guide, 5th edn. Muthen and Muthen, Los Angeles, CA (2007)
18. R Development Core Team: R: A language and environment for statistical computing. R Foundation for Statistical Computing, Vienna, Austria (2009). URL `http://www.R-project.org`. ISBN 3-900051-07-0
19. Robbins, N.B.: Creating More Effective Graphs. John Wiley & Sons, Hoboken (2005)
20. Schiffman, H.R.: Constancy and illusion. In: Sensation and perception, 5th edn., pp. 250–286. Wiley, New York (2002)
21. Schmid, C.F.: Statistical Graphics: Design Principles and Practices. Wiley, New York (1983)
22. Tufte, E.R.: The Visual Display of Quantitative Information. Graphic Press, Cheshire (2001)
23. Tukey, J.W.: Exploratory data analysis. Addison-Wesley, Reading, MA (1977)
24. Wainer, H.: Visual Revelations: graphic tales of fate and deception from Napoleon Bonaparte to Ross Perot. Copernicus, New York (1997)
25. Wright, D.B., London, K.: Modern Regression Techniques Using R. Sage, Washington DC (2009)

Chapter 6
Combinatorial Fusion for Improving Portfolio Performance

H. D. Vinod, D. F. Hsu and Y. Tian

Abstract A central issue for managers or investors in portfolio management of assets is to select the assets to be included and to predict the value of the portfolio, given a variety of historical and concurrent information regarding each asset in the portfolio. There exist several criteria or "models" to predict asset returns whose success depends on unknown form (parameters) of underlying probability distributions of assets, and whether one encounters a bull, bear or flat market. Different models focus on different aspects of historical market data. We use the recently developed Combinatorial Fusion Analysis (CFA) in computer science to enhance portfolio performance and demonstrate with an example using U.S. stock market data that fusion methods can indeed improve the portfolio performance. The R software is found to offer powerful tools for application of CFA in Finance.

Key words: rank-score function; combinatorial fusion analysis (CFA); stock performance; return on equity
JEL Classification Codes: G11; C14; D81

H. D. Vinod
Department of Economics, Fordham University, Bronx, NY 10458, USA
e-mail: vinod@fordham.edu

D. F. Hsu
Department of Computer & Information Science, Fordham University, New York, NY 10023, USA e-mail: hsu@cis.fordham.edu

Y. Tian
Department of Applied Mathematics and Computational Science, University of Pennsylvania, Philadelphia, PA 19104, USA e-mail: ytian001@hotmail.com

H.D. Vinod (ed.), *Advances in Social Science Research Using R*,
Lecture Notes in Statistics 196, DOI 10.1007/978-1-4419-1764-5_6,

6.1 Introduction

In managing a portfolio system, investors (or managers) aim to assemble a portfolio which can achieve the highest possible (optimal) risk-adjusted return. However, perfect optimality is an elusive goal in the uncertain world of asset markets based on past data, since the past data cannot reveal what the future might hold. Based on information such as historical performance of each of the assets, the investor uses different criteria or models to select assets to be included in the portfolio. A large number N of criteria have been used such as: price to earning ratio (PE), earnings per share (EPS), price to book value ratio (PBV), net margin (NM), net income to net revenue ratio (NINR), cash flow per share (CFS) and many others, often listed at financial websites. There are strong supporters and critics for each criterion, and none of them work under all market sentiments and economic conditions.

The two most popular models for portfolio management are the Capital Asset Pricing Model (CAPM) and the Arbitrage Pricing Theory (APT). Both rely on the mean–variance interrelationship among the assets in the portfolio. It is possible to incorporate utility theory into risk management; some nonlinear structures such as neural networks have also been used to forecast returns (Vinod and Reagle [10]). If the market uncertainty could be characterized by the bell-shaped normal distribution, mean–variance models do indeed yield optimal portfolios.

If not bell-shaped, the Pearson family of distributions can yield a very large variety of shapes based on a handful of parameters. More generally, there are lognormal, inverse-Gaussian, Azzalini skew-normal, Pareto-Levy-type stable distributions. Again, if one knew the correct parameters of the correct probability distribution describing the future market, the optimal portfolio can be obtained (Vinod and Reagle [10]). Unfortunately, there remains uncertainty regarding the choice of the correct distribution. Asset market professionals often distinguish between bull, bear and flat market sentiment for various assets at various times, noting that different kinds of uncertainty (probability distributions) apply for bull versus bear markets, while they obsess about the switch from a bull to bear market and vice versa.

Even if we know the market sentiment and the right probability distribution, one needs to contend with estimation uncertainty (see Vinod and Reagle [10], Vinod and Morey [8, 9]) about the parameters of the probability distribution based on limited historical data subject to measurement errors.

Thus, the "information" contained in historical market data is difficult to use as the number of model choices and underlying uncertainty increases. Let us view the stock market data as huge and diverse, ready to be exploratory "mined." This paper abandons the search for optimality and views portfolio choice as the one revealed by mining the data by using the Combinatorial Fusion Algorithm (CFA; Hsu, Chung and Kristal [3]).

The CFA begins with a set of multiple criteria, each of which implies a performance score. First, one reduces N, the total number of extant criteria, into a manageable size n. Then the systems are combined using a mathematical combinatorial algorithm, where $2^n - 1 - n$ "rank combinations" and $2^n - 1 - n$ "score combinations" are considered. Our method differs from other combination methods, e.g., those stated in Hazarika and Taylor [2], in many aspects including: our use of $2(2^n - n - 1)$ possible combinations, and our use of the concepts: rank-score function and "diversity." Section 6.2 describes CFA in the context of portfolio management following [11]. Section 6.3 describes our numerical experiment including the data set, the criteria used. The numerical result details and a discussion of possible future work are omitted for brevity. They are available in an electronic version with R software.

6.2 Combinatorial Fusion Analysis for Portfolios

A stock portfolio selects assets from thousands of stocks from the set $A = \{a_i\}$ of assets $a_1, a_2, \ldots$. Each such asset has a name, a ticker symbol, data on prices and performance based on its risk-adjusted returns. In this paper, the risk-adjusted return is measured by the ratio of return on equity (ROE) to the standard deviation (sd) of returns, or (ROE/sd).

We also have a large number of criteria or models $M = \{M_j\}$, having individual elements $M_1, M_2, \ldots$. The models here have abbreviations PE, EPS, PBV, NM, NINR, CFS, etc. We assume that we have data on these criteria score values for each asset. For example, stock ticker "XYZ" has some score for price earnings ratio, earnings per share, etc. In general, the numerical score of the i-th asset under j-th model is available for all assets and all models.

Since none of the individual stock picking criteria from PE, EPS, PBV, NM, NINR, CFS, etc. have been found to dominate others at all times, combinatorial fusion analysis combines them to define new fused criteria. However, there are two practical problems associated with combining these diverse criteria in their original form.

(i) The units of measurement for the scores are not comparable.

(ii) The original scores lack monotonic similarity in the sense that a stock with a *lower* PE ratio is more desirable, whereas a stock with a *higher* earnings per share is more desirable to be included in our portfolio.

We solve the first problem by mapping all original scores x to the unit interval $[0,1]$. That is, we use normalized (rescaled) values of x defined by

$$LO = min(x), UP = max(x), y = (x - LO)/(UP - LO). \tag{6.1}$$

We solve the second monotonic dissimilarity problem by simply changing the signs of all scores where less is desirable (e.g., PE ratio) to negative values.

Let these normalized sign-corrected scores be denoted by $\{s_{i,j}\}$. We can fix our focus on the j-th model M_j, sort these scores from the largest to the smallest and assign ranks starting with 1 for the most desirable, rank 2 for the next one, and so on. Thus, it is a simple matter to assign a numerical rank to the i-th asset under j-th model. Let all such ranks be denoted by $\{r_{i,j}\}$.

Now these normalized sign-corrected scores and ranks are ready to be combined (fused) by pairs, triplets, etc. It is convenient to begin with a discussion of combination by pairs. For illustration, let us combine the first two models M_1 and M_2 (PE and EPS, say) with the i-th asset scoring $s_{i,1}$ and $s_{i,2}$, respectively. Their score combination (SC) is defined for each i-th asset as the simple average of the two scores: $SC_{i,1+2} = 0.5(s_{i,1} + s_{i,2})$, where the subscript $1+2$ denotes the combination of M_1 and M_2 by averaging (not adding). A similar rank combination is also defined for the i-th asset as: $RC_{i,1+2} = 0.5(r_{i,1} + r_{i,2})$. A finance professional will have serious qualms about combining PE and EPS into one criterion. A computer scientist can abstract from the underlying names, willy-nilly combine them into one criterion and proceed to sort all assets by $SC_{i,1+2}$ values. One wants to identify assets with the highest score. This paper shows that such abstraction is potentially profitable.

First assume that we focus on an abridged set of p models $M_1, M_2, \ldots, M_p$. Hsu and coauthors [3] and [1] describe several ways of combination. In this paper, we use only the average combination, because our emphasis here is more on comparing rank and score combinations as in Hsu and Taksa [4]. For the set of p models $M_1, M_2, \ldots, M_p$, we define the score function of the score combined model SC as

$$SC_{i,1+\ldots+p} = \left(\sum_{j=1}^{p} s_{i,j} \right) / p. \tag{6.2}$$

Sorting the array $SC_{i,1+\ldots+p}$ into decreasing order would give rise to the rank function of the score combined model SC, written as r_{SC}. Similarly, we define the score function of the rank combined model RC as

$$RC_{i,1+\ldots+p} = \left(\sum_{j=1}^{p} r_{i,j} \right) / p. \tag{6.3}$$

Sorting this array $SC_{i,1+\ldots+p}$ into increasing order gives rise to the rank functions of the rank combined model RC, written as r_{RC}.

For each criterion M_j let $P(M_j)$ be the performance of M_j. We are most interested in the combination subset $C^{(j)}$ where $C^{(j)} \subset \{M_1, M_2, \ldots M_p\}$ so that $P(C^{(j)}) \geq \max_j P(M_j)$. We will call these **positive cases**. If $P(C^{(j)}) > \max_j P(M_j)$, we will call these **strictly positive cases.** If $P(C^{(j)}) < \max_j P(M_j)$, we will call these **strictly negative cases.** Obviously, our approach will suggest right stocks to buy if we find portfolio combinations leading to definitive performance improvements as revealed by strictly positive cases. Note that strictly negative combinations indicate stocks worth selling and do remain of interest.

Combinatorial fusion analysis has been used in information retrieval and virtual screening (see Hsu and Taska [4], Ng and Kantor [5] and Yang et al. [12]) with several applications in natural sciences. The framework of CFA and a survey is given in Hsu, Chung and Kristal [3]. Certain experience-based observations in the field of CFA are that combinations improve the performance only if: (a) individual systems have relatively good performance and (b) individual systems are diverse.

A rank-score function is defined by $f_M : N = \{1, 2, \ldots |A|\} \rightarrow [0, 1]$, where A is a set of all stocks and $|A|$ is the cardinality of the set A. Thus, we write

$$f_M(i) = s_M(r_M^{-1}(i)) = (s_M \circ r_M^{-1})(i), \tag{6.4}$$

where i denotes the rank.

The graph of the rank-score function f_M is the graph f_M with rank as the x-coordinate and score as the y-coordinate. See Fig. 6.1.

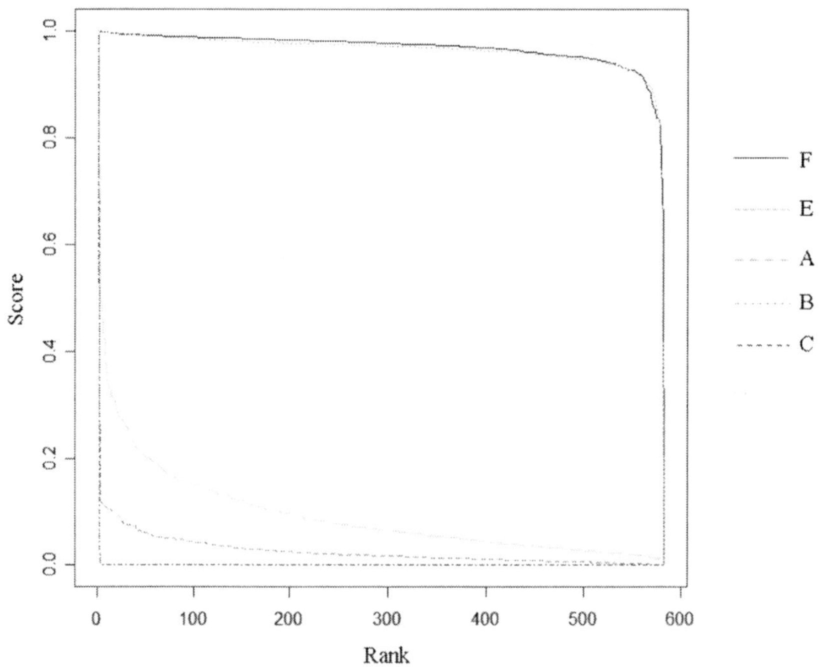

Figure 1(a) plot of rank-score functions (A,B,C,E,F)

Fig. 6.1 Plot of rank-score functions (A,B,C,E,F).

The diversity between systems M_1 and M_2, $d(M_1, M_2)$ either:

(1) $d(M_1, M_2) = d(s_{M1}, s_{M2})$ = product moment correlation between s_{M1} and s_{M2},

(2) $d(M_1, M_2) = d(r_{M1}, r_{M2})$ = rank correlation, [Spearman's ρ (rho) or Kendall's τ (tau)] between r_{M1} and r_{M2}, or

(3) $d(M_1,M_2) = d(f_{M1},f_{M2})$, where f_M is the rank-score function, and d is a measure of distance.

Yang et al. [12] used the Euclidean distance for their d:

$$d(f_{M1},f_{M2}) = \left[\sum_{i=1}^{n}(1/n)\left[f_{M1}(i) - f_{M2}(i)\right]^2\right]^{1/2}. \quad (6.5)$$

Let $P(A)$ denote the performance of criterion A. The pairwise performance ratio of low to high is defined as

$$PR(A,B) = Pl/Ph = \frac{min\{P(A),P(B)\}}{max\{P(A),P(B)\}}. \quad (6.6)$$

A graphical insight is gained in this literature by a diversity–performance graph, which plots the performance ratio Pl/Ph on the horizontal axis and suitably defined pairwise diversity on the vertical axis. The strictly positive cases where fusions lead to strictly superior performance are indicated by circles (∘) on the graph and negative cases indicated by (x) graphic symbols. Past experience and experiments suggest that circles are usually toward the northeast area of the diversity–performance graph and x's are found in the southwest area. See Fig. 6.2.

6.3 An Illustrative Example as an Experiment

Admittedly, we do not expect universal agreement on the choice of ROE/sd as the performance criterion used here. In all, we have nine models or criteria of interest, but plan to combine only $p = 5$ out of $n = 9$ at a time. This means we must compare the performance of 126 groups of (9 choose 5) possible choices of 5 out of 9. The notion of groups is new in this paper.

Our algorithm tries to focus on criteria making sure that they are all individually high performers. We have $2^5 - 1 - 5 = 26$ rank combinations and 26 score combinations of up to five models. We explain the insights from rank–score function and diversity-performance graphs. The details of the algorithm using the open source R package for this purpose are given in Sect. 6.3.2. It was implemented almost immediately on a PC with a 2-GHz processor.

6.3.1 Description of the Data Set

Our data are from Prof. Aswath Damodaran's website at the Stern Business School of New York University: `http://pages.stern.nyu.edu/~adamodar/New_Home_Page/data.html` The original data source is Value Line Inc. The

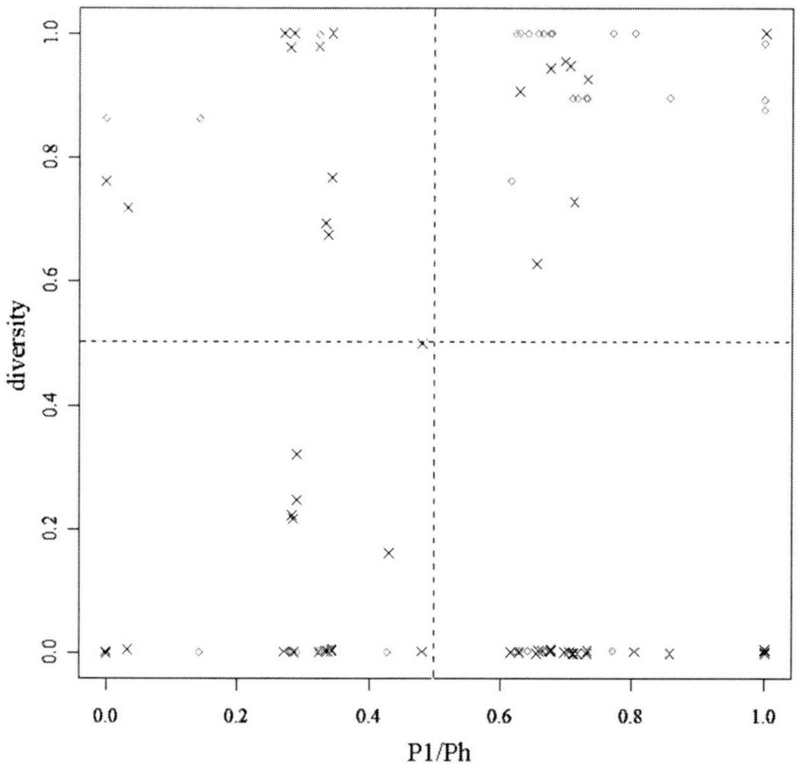

Fig. 6.2 Diversity–performance graph. Figure 3

site reports data for 7113 stocks (identified by ticker symbols and row numbers) along 7113 rows of an EXCEL workbook having 72 columns. The workbook columns are potential stock selection criteria involving the usual financial statistics including the PE ratio obtained by dividing the company's share price by its earnings per share (EPS), or price to book value (PBV) ratio as the ratio of market value of equity to book value of equity. The PBV is a measure of shareholders' equity in the balance sheet of a company. For our illustration we select data from the following nine criteria with following names and associated symbolic abbreviations used in our discussion below:

[A]=Trailing PE, [B]=Forward EPS, [C]=Forward PE, [D]=PBV Ratio, [E]=Ratio of Enterprise Value (EV) to Invested Capital, [F]=Value to B.V. of Capital, [G]=Growth in EPS during the last 5 years, [H]=Growth in Revenue last year, and finally, [K]=Net Margin.

First we construct an abridged data matrix, placing the nine criteria along nine columns and the risk-adjusted performance (ROE/sd) as the tenth column. We clean out all those rows (remove stocks) from the workbook which have missing data ending up with only 1129 rows in the abridged workbook (of original 7113). Each of the 1129 stocks is a candidate for inclusion in our proposed portfolio.

In order to ensure that each of the multiple systems satisfies "monotonic similarity," that is, they satisfy the same increasing or decreasing norm, i.e., the bigger the better, we multiply values in these nine columns (A,...,H,K) by the vector c(-1,1,-1,-1,1,1,1,1,1), where (-1) means it is desirable to have smaller values. For example, since it is desirable to have a small price earnings ratio, we multiply the column for A=Trailing PE by -1, as indicated. Similarly we change the sign of all values in columns for C=Forward PE and D=PBV Ratio. In the end, we make sure that all columns have numbers so that bigger values are preferred by investors.

6.3.2 Description of the Steps in Our R Algorithm

(1) Defining scores and computing their ranks. The 1129 normalized to [0,1] range and sign-corrected values in the nine columns entitled (A,...,H,K) are vectors of "scores" achieved by that stock under that criterion. Next, we rank these values from the smallest to the largest. The largest numbered rank (1129) means the highest score is different from the computer science literature where rank 1 is the best score.

(2) Choosing the best 113 stocks implicitly recommended by each criterion. The best stocks for each criterion reside along the bottom 113 (=10%) rows for each criterion in terms of scores as well as ranks.

(3) Union of all potentially desirable stocks. Next step is to consider a union of the 113 stocks recommended by each of the nine criteria with possibly $113 \times 9 = 1017$ stocks. Of course, even if the criteria (A,...,H,K) are distinct, the same stocks are implicitly recommended (i.e., among the top 113) by more than one criteria. In our example the union contains $n = 584$ stocks and 433 repetitions.

(4) Procedure for the one-criterion-at-a-time case. We propose to buy the best-performing subset of these 584 stocks. Next, we construct a three-column matrix denoted for brevity as **C3** with three columns C_1, C_2 and C_3. C_1 has numbers 1 to 584, C_2 has either the score values or the ranks and C_3 has ROE/sd. We sort this entire **C3** matrix on the second column, so that the best stocks will again be at the bottom of the matrix. Now we assume that chosen portfolio contains the best 10% stocks by each criterion and compute the performance of the portfolio from the "average ROE/sd." A check on our programming is that the average ROE/sd should be exactly the same whether we use scores or ranks in C_2 of the matrix here, as long as we have only one criterion at a time, that is, before we combine them by a fusion algorithm.

In the traditional method of portfolio selection the computation can end here. It is, however, only the beginning under our proposal. Instead of being satisfied with choosing only one criterion at a time, we consider combinations of two or more criteria. For simplicity the combinations considered in this paper are simple averages, but weighted averages can be considered without loss of generality.

(5) Procedure for the two-criteria-at-a-time case. We combine two criteria at a time and compute performance for each pair. There are (9 choose 2) or 36 possible pairs of criteria from (A, B, C, D, E, F, G, H, K); for example, AB, AC, AD, AE, etc. For each such pair we create a matrix **C3** with three columns similar to Step 4. We sort entire matrix **C3** on the second column (best at the bottom). Now we select the portfolio of the best 10% stocks for each paired criterion and compute the average ROE/sd for these chosen stocks. It is perhaps not obvious that, unlike the one-at-a-time case above, the ROE/sd numbers (hence recommended stocks) are different when the second column of **C3** contains average scores instead of ranks.

Compared to the traditional method of choosing one criterion at a time, the fusion algorithm is ahead of the game if we have the strictly higher ROE/sd for any combination of two criteria at a time. For our example, the combination of criteria A and E is often found to be superior to A or E alone.

(6) Procedure for the general k-criteria-at-a-time case. If we have k=3,4,5 criteria at a time we must enumerate all (9 choose k)=(84, 126, 126) choices similar to ABC, ACD, ADE, etc. For each k-at-a-time fusion set we again create and sort the entire matrix **C3** with three columns as before on the second column (having ranks or scores yielding the best at the bottom) and make a portfolio of the best 10% stocks for each k-at-a-time fusion set and compute the performance of the portfolio by the average ROE/sd.

After finishing this for k=1 to 5 we will have $2 \times 381 = 762$ performance numbers for each stock associated with the nine criteria (A to H and K) and their fusions involving k at a time, where the doubling is needed because we have score-based numbers as well as rank-based numbers. Of the 762, we can ignore the k=1 case leading to 744 relevant ROE/sd numbers to be compared.

(7) The total of 126 groups. Even if we have 9 criteria we have determined that it is impractical to use a criterion for the choice of stocks based on a fusion of more than five stock-picking criteria at a time in our context. This means we are not allowing a grand fusion of all 9 criteria (ABCDEFGHK). We are also disallowing fusions containing 6, 7, or 8 criteria at a time. However, we must consider a complete listing of choices for all possible sets of 5 out of 9 leading to (9 choose 5) or 126 distinct choices to be considered separately. We refer to these as groups for the purpose of discussion. Since the portfolio of best stocks recommended for one group will not, in general, coincide with the best portfolio for another group, we need to study all of them.

(8) Final ranking of all stock choices for all 126 groups. Compared to the traditional method of choosing one criterion at a time, the fusion method is much ahead of the game, since we have noticed that a great many cases exist, where combined criteria using averages of scaled scores have strictly higher ROE/sd performance than the ROE/sd of their individual components taken one at a time. Now we will consider $2 \times (2^5 - 1 - 5) = 52$ ROE/sd numbers for each of the 126 groups. However, we fully expect to have many duplicate fusions among these groups. It turns out that we need not be concerned with explicitly separating the duplicates, because we can simply rank order with respect to $52 \times 126 = 6552$ ROE/sd numbers from the lowest ROE/sd to the largest ROE/sd, and eventually pick the best rows based on the highest ROE/sd. If there are duplicates they will simply become identical rows, easily omitted by a computer algorithm. Identical rows do not affect the value of ROE/sd or the ranking.

For our example, the best fusion is the rank combination of criteria A, B, E, and F. The electronic version having R software lists ticker symbols for the top 50 stocks recommended by this ABEF combination and the algorithm in far greater detail. All plots produced by R are not included here for brevity. We note that R is particularly powerful in our context of combinatorial fusion to construct portfolios to buy.

By way of extension, it is also quite possible to construct stock portfolios to sell, use time series data for each stock and myriad other choices of performance and stock-picking criteria. For example, Vinod [6] discusses four orders of stochastic dominance based on different empirical probability distributions suggested by the past data for each stock. With the use of the fusion algorithm we are not restricted to focus on only one stochastic order of dominance at a time. Vinod [7] discusses the statistical theory behind CFA. We hope we have convinced the reader that the CFA approach can have a significant potential and an attractive future in practical processes, as well as the art and science of portfolio selection.

References

1. Chung, Y.-S., Hsu, D.F., Tang, C.Y.: On the diversity–performance relationship for majority voting in classifier ensembles. In: Multiple Classifier Systems, MCS, LNCS No. 4472, pp. 407–420. Springer (2007)
2. Hazarika, N., Taylor, J.G.: Combining models. In: Proceedings of the 2001 International Joint Conference on Neural Networks, IJCNN-2001, pp. 1847–1851. Washington, DC (2001)
3. Hsu, D.F., Chung, Y.-S., Kristal, B.S.: Combinatorial fusion analysis: Methods and practices of combining multiple scoring systems. In: Advanced Data Mining Technologies in Bioinformatics, pp. 32–62. Ideal Group, Inc. (2006)
4. Hsu, D.F., Taksa, I.: Comparing rank and score combination methods for data fusion in information retrieval. Information Retrieval **8**, 449–480 (2005)

5. Ng, K.B., Kantor, P.B.: Predicting the effectiveness of naïve data fusion on the basis of system characteristics. Journal of the American Society for Information Science **51**(13), 1177–1189 (2000)
6. Vinod, H.D.: Ranking mutual funds using unconventional utility theory and stochastic dominance. Journal of Empirical Finance **11**(3), 353–377 (2004)
7. Vinod, H.D.: Ranking and selection among mutual funds. In: N. Balakrishnan (ed.) Methods and Applications of Statistics in Business, Finance and Management Sciences. John Wiley and Sons, Hoboken, New Jersey (2010). To appear
8. Vinod, H.D., Morey, M.: A double sharpe ratio. In: C.F. Lee (ed.) Advances in Investment Analysis and Portfolio Management, vol. 8, pp. 57–65. JAI-Elsevier Science (2001)
9. Vinod, H.D., Morey, M.: Estimation risk in Morningstar fund ratings. Journal of Investing **11**(4), 67–75 (2002)
10. Vinod, H.D., Reagle, D.P.: Preparing for the worst: Incorporating downside risk in stock market investments. Wiley Interscience (2005)
11. Vinod, H. D., Hsu, D. F., Tian, Y.: Combining multiple criterion systems for improving portfolio performance. Discussion paper no. 2008-07, Department of Economics, Fordham University (2008). URL `http://ssrn.com/abstract=1127879`
12. Yang, J.-M., Chen, Y.-F., Shen, T.-W., Kristal, B.S., Hsu, D.F.: Consensus scoring criteria for improving enrichment in virtual screening. Journal of Chemical Information and Modeling **45**, 1134–1146 (2005)

Chapter 7
Reference Growth Charts for Saudi Arabian Children and Adolescents

P. J. Foster and T. Kecojević

Abstract The purpose of this study is to provide Saudi Arabian population reference growth standards for height, weight, body mass index (BMI), head circumference and weight for length/stature. The estimated distribution centiles are obtained by splitting the population into two separate age groups: infants, birth to 36 months and children and adolescents, age 2 to 19 years. The reference values were derived from cross-sectional data applying the LMS method of Cole and Green (*Statistics in Medicine* 1992; **11**:1305–1319) using the `lmsqreg` package in `R` (public domain language for data analysis, 2009). The report provides an overview of how the method has been applied, more specifically how the relevant issues concerning the construction of the growth charts have been addressed, and is illustrated by just using the girls' weight data (birth to 3 years old). These issues include identifying the outliers, diagnosing the appropriate amounts of smoothing and averaging the reference standards for the overlapping 2- to 3-year age range. The use of ANCOVA has been introduced and illustrated as a tool for making growth standard comparisons between different geographical regions and between genders.

Key words: growth curves; quantile regression; LMS; cross-sectional data

---

P. J. Foster
School of Mathematics, University of Manchester, Manchester, UK
e-mail: `Peter.Foster@manchester.ac.uk`

T. Kecojević
Lancashire Business School, University of Central Lancashire, Preston, UK
e-mail: `TKecojevic@uclan.ac.uk`

Prepared for presentation at: *Conference on Quantitative Social Science Research Using R*, Fordham University, New York, June 18-19, 2009.

H.D. Vinod (ed.), *Advances in Social Science Research Using R*,
Lecture Notes in Statistics 196, DOI 10.1007/978-1-4419-1764-5_7,

7.1 Introduction

The growth standards are derived from a cross-sectional sample of healthy children and adolescents aged from birth to 19 years. The sample was randomly selected by a stratified multistage probability sampling procedure from each of the 13 administrative regions of the Kingdom of Saudi Arabia, ensuring both national and urban/rural representation. The anthropometric data comprises 51,485 observations of which 25,987 are made on boys and 25,498 on girls. Those measurements include: length, for the children 2 years of age and below, height, for children above 2 years of age, weight and head circumference. All possible efforts have been made to ensure reliability and the accuracy of the measurements.

The reference growth charts we have constructed describe the dependence of height, weight, body mass index (BMI) and head circumference on age, and weight on length/stature for two age ranges, birth to 36 months and 2 to 19 years. They were constructed using the LMS (Lamda-Mu-Sigma) method of Cole and Green [8] in `R`, a public domain language for data analysis (R Development Core Team (2009)). The LMS method provides a way of obtaining growth standards for healthy individuals and is based on normalizing the conditional distribution of a measure using the power transformation of Box and Cox [1]. The package `lmsqreg` developed by Carey [2] implements the LMS method in `R`. Use of the LMS method was a requirement of the study.

In this paper we discuss various issues involved in using the LMS methodology, all of which are specifically illustrated using the Saudi girls' weight data for those from birth to 3 years old. Section 7.2 focuses on identifying and removing extreme outliers prior to estimating the centile curves. In Section 7.3 we describe the model and how it is fitted to the data. Goodness-of-fit is considered in Sect. 7.4 while in Sect. 7.5 we describe a simple solution to obtaining a common set of centiles in the overlapping 2- to 3-year age range. In Section 7.6 we propose using ANCOVA to investigate differences in growth patterns in different geographical regions and also between the sexes. This proves to be a much more informative approach than that described in the literature which does not take age into account. Finally, we add some discussion and further suggestions.

7.2 Outliers

An *outlier* is a sample value that lies outside the main pattern or distribution of the data and in the context of quantile regression, which was first introduced by Koenker and Bassett [14], it will be one which has a much larger or smaller response value at a given age when compared with other responses at a similar age. Quantile regression measures the effect of covariates not only in the center of the distribution but also in the upper and lower tails. Extremely low and extremely upper quantiles are of interest regarding growth charts and therefore it is important to deal with the issue of removing the potential outliers with cautiousness. An outlier should not be regarded as a pejorative term; outliers may be correct, but they should be checked for transcription error [18]. The quantile regression model is a natural extension of the linear regression model. If an outlier is included in the data which is used to estimate the quantiles, then it may be highly influential on the fitted regression line in that the line may be pulled in a disproportionate manner towards the outlying value or it may cause a failure in the algorithm used to estimate the quantiles [14]. This latter point is particularly true with respect to the LMS procedure, as according to Carroll [4] the choice of the transformation L(x) is highly sensitive to outliers in the data. We have also found that if the outliers are not removed, it can result in the numerical failure of the model fitting algorithm in the function `lmsqreg`. The lack of a methodology to assess the direct effect of an individual observation on the LMS methodology has prompted us to approximate the LMS model using a cubic regression line to model the relationship between a response and covariate (such as weight and age). Approximating the LMS model in this way enables us to identify the outliers in that space with respect to this mode, that hopefully are also the outliers with respect to the LMS model. To fit this cubic regression line we have used a robust regression procedure. Robust regression deals with cases that have very high leverage, and cases that are outliers. Robust regression represents a compromise between the efficiency of the ordinary least squares (OLS) estimators and the resistance of the least absolute value (LAV) estimators, both of which can be seen as special cases of M-estimation [13]. It is a form of weighted least squares regression, which is similar to least squares in that it uses the same minimization of the sum of the squared residuals, but it is done iteratively. Based on the residuals a new set of weights are determined at each step. In general, the larger the residuals are, the smaller the weights. So the weights depend on the residuals. At the same time, the residuals depend on the model and the model depends on the weights. This generates an iterative process and it goes on until the change in the parameter estimates is below a preset threshold. At the end, instead of all points being weighted equally, the weights vary and those with the largest weights contribute more to the fit.

There are a few types of weighting schemes, M-estimators, that can be implemented [18]. In Huber's [13] weighting, observations with small residuals get a weight of 1; the larger the residual, the smaller the weight. M-estimation, introduced by Huber [12] can be regarded as a generalisation of maximum-likelihood estimation (MLE), hence the term 'M'-estimation [10].

Consider the linear model

$$y_i = x_i^{'}\beta + \varepsilon_i \qquad i = 1, \ldots, n \tag{7.1}$$

where the $V(\varepsilon_i) = \sigma^2$ and $Cov(\varepsilon_i, \varepsilon_j) = 0, i \neq j$. If ε_i has density f, we can define $\rho = -\log f$, where the function ρ gives the contribution of each residual to the objective function. Then the MLE $\hat{\beta} = b$ solves

$$\min_{\beta} \sum_{i} -\log f(y_i - \mu_i) = \min_{\beta} \sum_{i} \rho(y_i - \mu_i) \tag{7.2}$$

where $\mu_i = x_i^{'}\beta$ and so $\hat{\mu}_i = x_i^{'}b$.

Let $\psi = \rho^{'}$ be the derivative of ρ. Then we will have $\sum_i \psi(y_i - \hat{\mu}_i)x_i^{'} = 0$ or $\sum_i w_i(y_i - \hat{\mu}_i)x_i^{'} = 0$ where the weight $w_i = \psi(y_i - \hat{\mu}_i)/(y_i - \hat{\mu}_i)$. This suggests an iterative method of solution, updating the weights at each iteration [18].

If $\rho(x) = x^2$, the solution is the conditional mean and the median is $\rho(x) = |x|$. The function

$$\psi(x) = \begin{cases} -c & x < -c \\ x & |x| < c \\ c & x > c \end{cases} \tag{7.3}$$

is known as *Winsorizing* and brings in extreme observations to $\mu \pm c$. The corresponding function $\rho = -\log f$ is

$$\rho(x) = \begin{cases} x^2 & \text{if} |x| < c \\ c(2|x| - c) & \text{otherwise} \end{cases} \tag{7.4}$$

and equivalent to a density with a Gaussian centre and double-exponential tails. This estimator is due to Huber. Note that its limit as $c \to 0$ is the median, and as $c \to \infty$ the limit is the mean. The value c =1.345 gives 95% efficiency at the normal [18].

Venables and Ripley's `MASS` package [17] introduces the `rlm` function for fitting a linear model by iterated re-weighted least squares (IWLS) regression using Huber's M-estimator with tuning parameter c =1.345 and also incorporating a robust estimate of the scale parameter σ, where $\hat{\sigma} = s$. If we assume a scaled pdf $f(e/\sigma)/\sigma$ for ε and set $\rho = -\log f$, in this case the MLE minimizes

$$\min_{\beta} \left[\sum_{i} \rho\left(\frac{y_i - \mu_i}{\sigma}\right) + n\log\sigma \right] \tag{7.5}$$

Assuming that σ is known and if $\psi = \rho^{'}$, then the MLE b of β solves

$$\min_{\beta} \sum_{i} x_i \psi\left(\frac{y_i - \mu_i}{\sigma}\right) = 0 \tag{7.6}$$

A common way to solve the above equation is by IWLS, with weights

$$w_i = \psi\left(\frac{y_i - \hat{\mu}_i}{\sigma}\right) \Big/ \left(\frac{y_i - \hat{\mu}_i}{\sigma}\right) \tag{7.7}$$

Of course, in practice the scale σ is not known. However, as mentioned above σ is estimated by a robust MLE-type estimate denoted by s.

A cubic polynomial using the **rlm** function in **R** has been fitted to the log-transformed data (in a bid to stabilize the variance over age) using *MM*-estimation that combines the resistance and robustness, while gaining the efficiency of *M*-estimation.

```
> library(MASS)
> mp<-rlm(log(weight)~1+agey+I(agey^2)+I(agey^3), method="MM")
> summary(mp)

Call: rlm(formula = log(weight) ~ 1 + agey + I(agey^2) +
                                I(agey^3),  method = "MM")
Residuals:
      Min        1Q    Median        3Q       Max
-0.784423 -0.098632 -0.001707  0.096245  0.708137

Coefficients:
            Value    Std. Error t value
(Intercept)   1.1731   0.0034   342.0763
agey          2.0866   0.0148   140.8422
I(agey^2)    -1.1875   0.0145   -81.6688
I(agey^3)     0.2223   0.0036    61.0781

Residual standard error: 0.144 on 6123 degrees of freedom
```

After fitting this cubic line we have used the weights produced in a robust regression procedure to identify the most extreme values. The observations with the big residuals are down weighted, which reflects that they are atypical from the rest of the observations when it comes to fitting such a model. Observations with 0 weight ($w_i = 0$) are deemed to be extreme and so are then removed from the data before running the LMS model fitting algorithm (Figs. 7.1 and 7.2). Note that weight referred to in Fig. 7.1 corresponds to girls' actual body weight.

R Data Editor

row.names	weight	agey	w
811	15.7	0.6	0
1235	2.5	0.2617	0
2240	4.8	1.21	0
5963	30.2	2.96	0
1149	11	0.33	0.002548397
1993	6.8	2.7819	0.00273806
3250	20	1.86	0.02485549
4170	8	0.15	0.02486067
1201	9.8	0.3	0.06277497
1305	8.4	0.2	0.06360398
3454	14	0.65	0.1121188
86	18.5	1.97	0.1242005
4131	3	0.259	0.1309751
3595	7	0.1206	0.1326621
5740	6	1.0648	0.1556946
5392	10	0.36	0.1615647

Fig. 7.1 Identifying the outliers, girls' weight, age birth to 36 months.

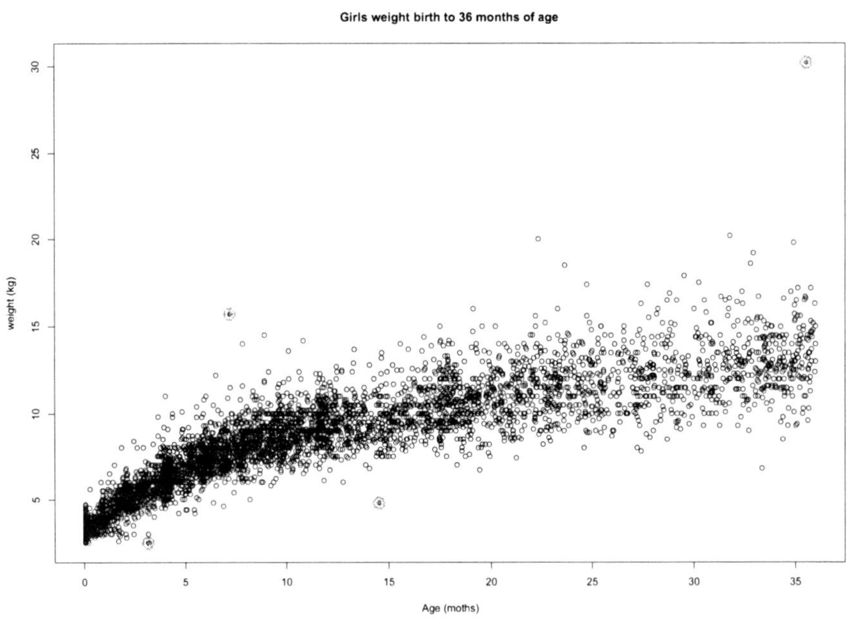

Fig. 7.2 Identified outliers, girls' weight, age birth to 36 months.

The World Health Organisation (WHO) has defined limits for acceptable data based on 1977NCHS/WHO growth charts and recommends that the exclusion range for weight-for-age should be $|z| > 5$ [5]. After the final LMS model for girls' weight (age birth to 3) was fitted, we used the `zscores` function from the `lmsqreg` package in `R` to calculate z-scores for the four identified outliers and these are given in Table 7.1. Each omitted case has an $|z|$ greater than 5 tying in with the WHO guideline.

Table 7.1 z-scores of the four identified outliers for girls' weight, age birth to 36 months

z-`scores {lmsqreg}`			
row.names	weight	age	z-score
811	15.7	0.6	5.22488
1235	2.5	0.2617	-6.70038
2240	4.8	1.21	-5.29738
5963	30.2	2.96	6.49793

7.3 LMS

Under the assumption of normality, growth curves can be constructed by estimating the age-specific mean and standard deviation, say $\mu(t)$ and $\sigma(t)$, so that chosen quantile curve for $\alpha \in [0,1]$ can then be obtained as

$$\hat{Q}(\alpha \mid t) = \hat{\mu}(t) + \hat{\sigma}(t)\Phi^{-1}(\alpha) \tag{7.8}$$

where $\Phi^{-1}(\alpha)$ denotes the inverse of the standard normal distribution function. Providing that the assumption of normality holds at each age, such a curve should split the population into two parts with the proportion α lying below the curve, and the proportion $1-\alpha$ lying above the obtained curve [19].

Although adult heights in a reasonably homogeneous population are known to be quite close to normal, in general anthropometric data are known to be not normally distributed [19]. Anthropometry tends to be right skew rather than left skew, which is why a log transformation which treats the two tails of the distribution differently is often suggested as a means of obtaining a symmetric distribution [7]. A log transformation can be viewed as a particular power transformation of the data but there is a whole family of such powers. Cole [6] suggested that in principle, there is no reason why a general power transformation should not be applied to the data. The maximum likelihood estimate (MLE) for the power, which both minimises the skewness and optimises the fit to normality, is ideally suited to the problem of skew data. However, it only operates on individual groups and does not allow for the skewness to change in a smooth manner over the range of the covariate.

The LMS, or $\lambda\mu\sigma$, approach of Cole [6] provides a way of obtaining normalised growth centiles that deals quite generally with skewness as well as nonconstant variance. The method enables us to fit the growth standards to all forms of anthropometry by making the simple assumption that the data can be normalised by using a smoothly varying Box–Cox transformation, so that after the transformation of the measurements $Y(t)$ to their standardised values $Z(t)$ they will be normally distributed:

$$Z(t) = \frac{[Y(t)/\mu(t)]^{\lambda(t)} - 1}{\lambda(t)\sigma(t)} \tag{7.9}$$

With these normalised measurements, the desired quantile curve for $\alpha \in [0,1]$ can then be obtained using the following model:

$$Q(\alpha \mid t) = \mu(t)[1 + \lambda(t)\sigma(t)\Phi^{-1}(\alpha)]^{1/\lambda(t)} \tag{7.10}$$

which summarises the construction of the centiles by three smooth curves, i.e., functions, representing the skewness, the median and the coefficient of variation. The LMS method works with power transformed measurements, but converts the mean back to original units and uses coefficient of variation (CV) rather than standard deviation of the data. In this way the results for different power transformations can be compared, and the best (Box–Cox) power can be identified as the one which gives the smallest CV [7]. This method provides a coherent set of smoothed centiles and the shape of the power curves provide information about the changing skewness, median and coefficient of variation of the distribution.

The three parameters λ, μ and σ were assumed to change smoothly with age. Green [11] has proposed to estimate the three curves by maximizing the penalised likelihood,

$$\ell(\lambda,\mu,\sigma) - \nu_\lambda \int (\lambda''(t))^2 dt - \nu_\mu \int (\mu''(t))^2 dt - \nu_\sigma \int (\sigma''(t))^2 dt \tag{7.11}$$

where $\ell(\lambda,\mu,\sigma)$ is the Box–Cox log-likelihood function derived from (7.9),

$$\ell(\lambda,\mu,\sigma) = \sum_{i=0}^{n} [\lambda(t_i) \log \frac{Y(t_i)}{\mu(t_i)} - \log\sigma(t_i) - \frac{1}{2} Z^2(t_i)] \tag{7.12}$$

and $Z(t_i)$ are the SD scores corresponding to $Y(t_i)$. In this way, the three curves are constrained to change smoothly as the covariate changes and, like the centiles, they can be plotted against the covariate (Figs. 7.3 and 7.4). The curves are fitted using cubic splines to give a nonlinear regression, and the extent of the smoothing required can be expressed in the terms of smoothing parameters $(\nu_\lambda, \nu_\mu, \nu_\sigma)$. These quantities are defined to be the traces of the relevant smoothing matrices and are referred to as the "equivalent degrees of freedom" (edf) [19]. Cole and Green [8] argued that the distributions of $(\nu_\lambda, \nu_\mu, \nu_\sigma)$ in the LMS model are largely independent of each other, implying that one edf can be optimised while fixing the other two.

```
> mw3<-lmsqreg.fit(weight, age, edf=c(7,13,9),
   pvec = c(0.03, 0.05, 0.1, 0.25, 0.5, 0.75, 0.9, 0.95, 0.97))
> plot(mw3)
> points(age, weight, pch=".",col="red")
```

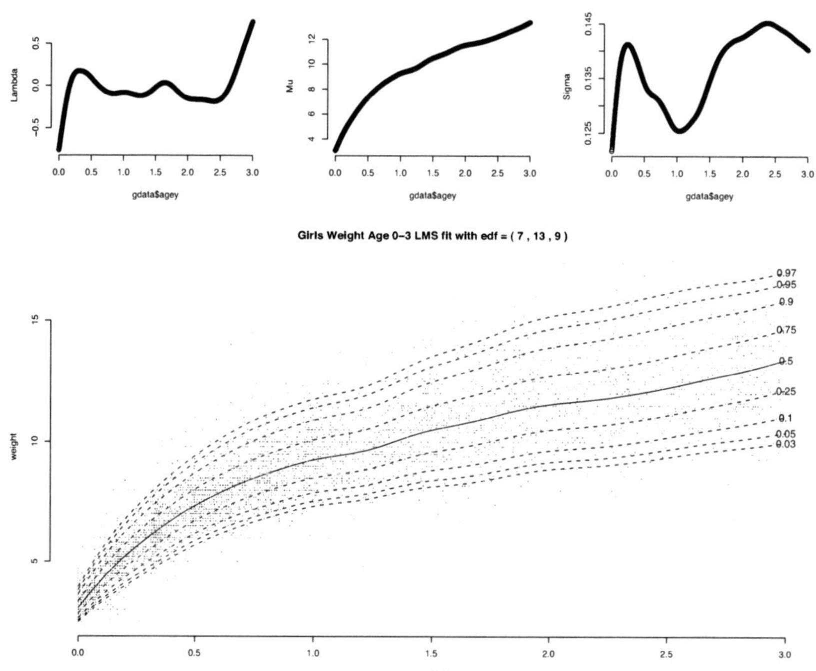

Fig. 7.3 Centile curves for girls' weight birth to 36 months of age.

Carey [2] has developed the `lmsqreg` package that implements the LMS method in R. Smoothed centile curves have been fitted to the reference data using the `lmsqreg.fit` function with suggested starting edf values setting of 3, 5 and 3 for λ, μ and σ, respectively [3]. The strategy is then to optimise the μ curve edf, by increasing/decreasing the edf by 1 until the change in penalised likelihood is small, i.e., less than 2. Once the μ curve is fitted, the process is repeated or the σ curve avoiding the value for edf of 2 which would force a linear trend on the μ curve. Finally, the λ curve was fitted similarly to the σ curve (Fig. 7.3). However, in cases of fitting the centile curves for weight measurement age 2 to 19 years for both sexes λ had to be set to a value of zero, which constrains the entire curve to be a constant value and forces a log transformation (Fig. 7.4). The same had to be applied for the fitting of male head circumference age 2 to 19 years.

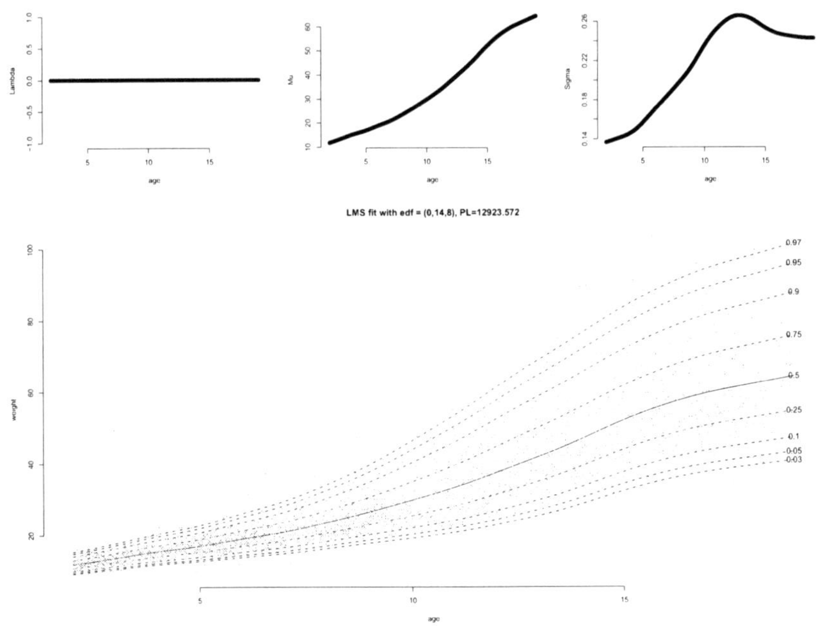

Fig. 7.4 Centile curves for boys' weight 2 to 19 years of age.

```
> mw19<-lmsqreg.fit(weight, age, edf=c(0,14,8), pvec = c(0.03,
   0.05, 0.1, 0.25, 0.5, 0.75, 0.9, 0.95, 0.97), lam.fixed=0)
> plot(mw19)
> points(age, weight, pch=".", col="cyan")
```

Following the suggested strategy, the data were overfitted and the curves were clearly undersmoothed. As Cole [8] implies, the case for making the centile curves smooth is to some extent cosmetic — the centiles are more pleasing to the eye when smoothed appropriately but it is also in the belief that the true population centiles will themselves change smoothly. Any nonparametric curve estimation method requires some means of controlling the smoothness of the fitted functions. For the LMS method this control is provided by the edf parameters $(\nu_\lambda, \nu_\mu, \nu_\sigma)$.

As indicated by Carey [2] the value by which to increase/decrease edf and the change in penalised likelihood depends on the sample size. For large samples the change of less than 2 units is not significant; therefore, the large change is needed and the final decision should depend on the appearance of the curve. In order to overcome the overfitting of the curves the edf values had to be relaxed.

7.4 Smoothing and Evaluation

The number of effective degrees of freedom is a convenient parameter that expresses the amount of adjustment necessary for smoothing a set of data. Adjustment of edf values was done following Carey's [2] algorithm, this time decreasing the value for ν_μ by 1 until the curve appeared to be smooth. The same procedure was followed for ν_σ and lastly for ν_λ (Fig. 7.5). Finally, the adequacy of the chosen model is evaluated using the original data.

As discussed by Green [11], the distribution theory for model evaluation statistics formed on the bases of changes in penalised likelihood is currently still undeveloped. We have adopted a local-test based approach to formal model evaluation. Carey's *lmsqreg* package [2] provides as a part of the output for a fitted model a collection of model-based z-scores derived from the given quantile regression model. They are stratified based on the covariate t, and within this strata, z-scores are tested for marginal Gaussianity (Kolmogorov–Smirnov test), zero mean (Student's t-test) and unit variance (χ^2 test) [3].

```
> mw3
Dependent variable: gdata$weight , independent variable: gdata$agey
The fit converged with EDF=( 4,6,3 ), PL= 9198.316

KS tests: (intervals in gdata$agey //p-values)
    (-0.001,0]      (0,0.348] (0.348,0.802]  (0.802,1.54]       (1.54,3]     Overall
         0.000          0.000         0.271         0.324          0.676       0.001

t tests: (intervals in gdata$agey //p-values)
    (-0.001,0]      (0,0.348] (0.348,0.802]  (0.802,1.54]       (1.54,3]     Overall
         0.006          0.000         0.562         0.369          0.568       0.810

X2 tests (unit variance): (intervals in gdata$agey //p-values)
    (-0.001,0]      (0,0.348] (0.348,0.802]  (0.802,1.54]       (1.54,3]     Overall
         0.000          0.000         0.717         0.050          0.462       0.979
```

The above output from the final fitted model shows that the hypotheses of a zero mean, unit variance normal distribution in the intervals close to birth are rejected. The original data are strongly skewed and the edf parameters finally selected are not able to transform the data sufficiently well, with the final empirical distribution being slightly skewed. If the smoothing parameters are increased, in particular ν_λ, the normality of the transformed data can be successfully achieved. However, as discussed earlier in Sect. 7.3, we reduced the values of the optimal smoothing parameters in order to obtain smoother estimated centile curves.

Table 7.2 reports on the accuracy of the quantile regression fit in terms of the discrepency between the nominal and empirical proportions of data lying beneath selected quantile function for age group birth to 3 years. By and large these results show that the quantiles of the fitted models do fit the data well.

Table 7.2 Table entries are quantile coverage probability estimates. Measurement: Age: birth to 36 months

			p								
sex	variable	N	0.03	0.05	0.10	0.25	0.50	0.75	0.90	0.95	0.97
female	weight	6,123	0.025	0.052	0.090	0.240	0.506	0.755	0.905	0.950	0.972

7.5 Averaging

We were required to produce reference standards for two age groups: birth to 36 months of age and 2 to 19 years of age. The overlap for the two sets of charts occurs for ages between 2 and 3 years. The values for both sets of standards in the overlapping age range are a product of the model fitted to the whole data set for each specific age group. This means that the centile curves for a particular measurement in this overlapping period will not be the same for the two sets of charts as they are based on using different data outside the range 2 to 3 years (Fig. 7.6).

One of the arguments of the `lmsqreg.fit` function is `targlen` which defines the number of points at which smooth estimates of λ, μ, and σ should be extracted for quantile plotting. For both sets of charts we have adopted the default value of 50 for the `targlen` argument. For the overlapping period 2 to 3 years this produces 17 points in the birth to 36 month chart and 3 points in the chart for ages 2 to 19 years (Fig. 7.7).

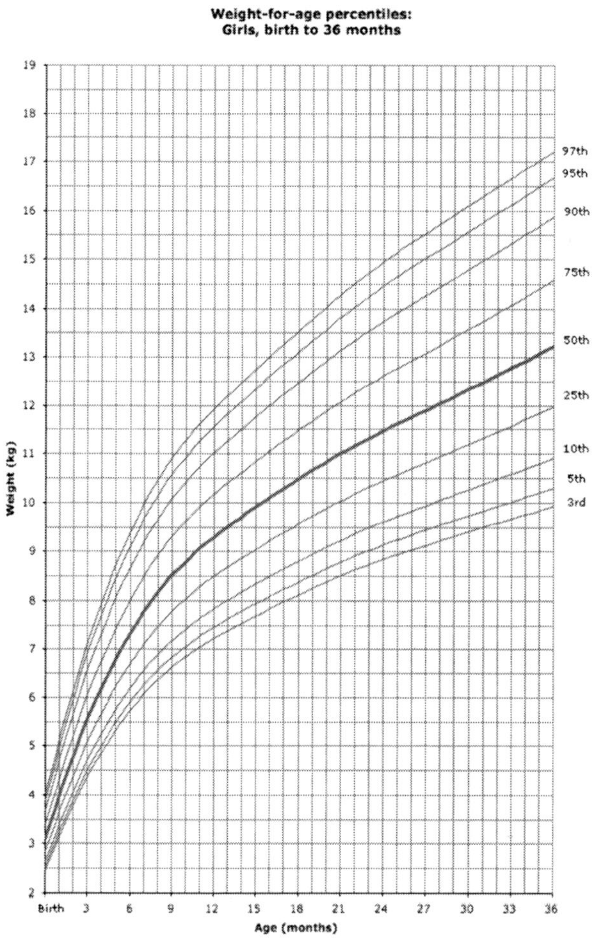

Fig. 7.5 Final smooth centile curves for girls' weight birth to 36 months of age.

In order to make the centile curves for a particular measurement for this overlapping period the same for the two sets of charts we have re-estimated the curves using the following cubic polynomial:

$$\hat{y}_i = \beta_0 + \beta_1 x_i + \beta_2 x_i^2 + \beta_3 x_i^3 \tag{7.13}$$

To estimate this cubic polynomial for each of the centiles at the lower and upper boundaries of the overlapping period we have used three adjacent points from each of the charts (Figs. 7.7 and 7.8), using the least squares estimator given by

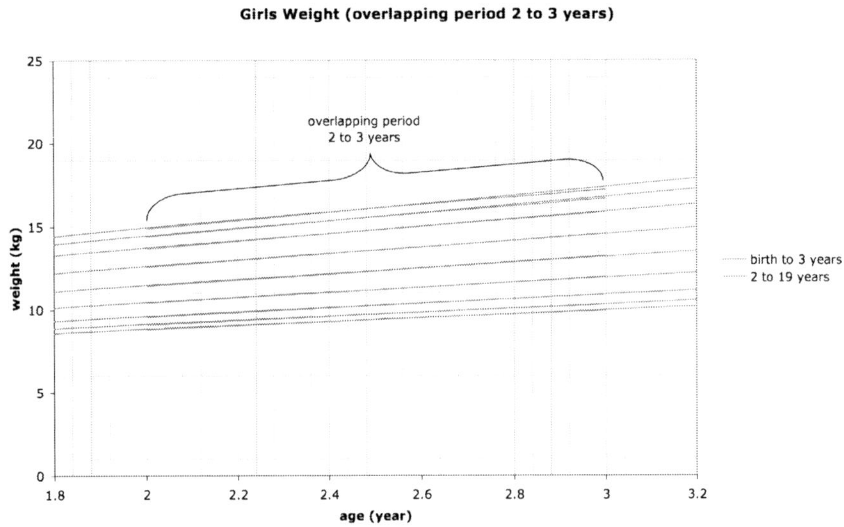

Fig. 7.6 Overlapping charts: girls' weight.

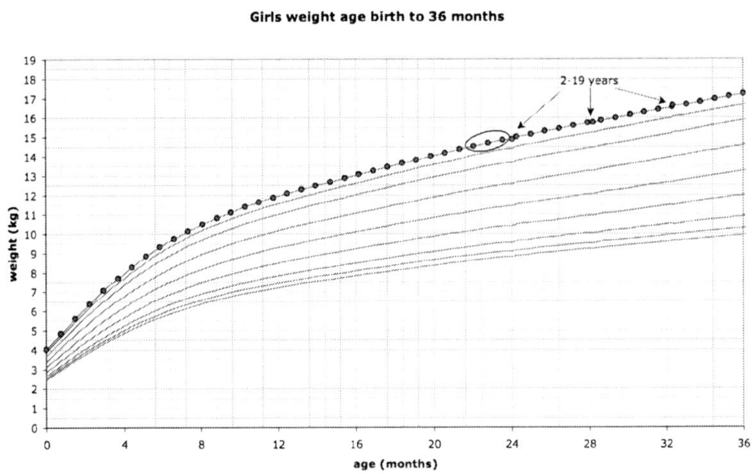

Fig. 7.7 Centile curves for girls' weight birth to 36 months of age.

$$\hat{Y} = X\left[X'X\right]^{-1}X'Y \tag{7.14}$$

For the overlapping period new estimates were calculated using the newly found polynomial resulting in a smooth overlap (Fig. 7.9). This means that the centiles for a particular measurement will be the same in the birth to 36 month chart as in the 2 to 19 year age chart.

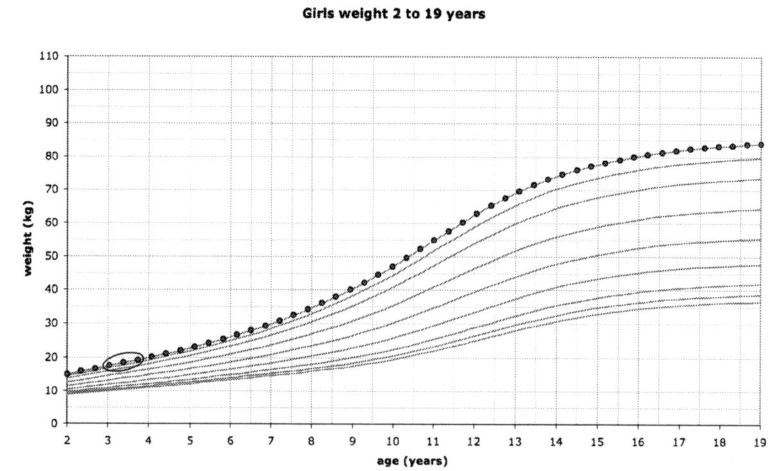

Fig. 7.8 Centile curves for girls' weight age 2 to 19 years.

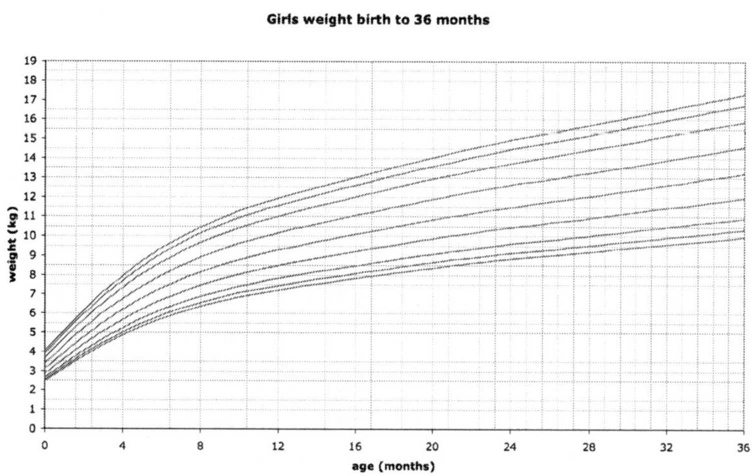

Fig. 7.9 Final smooth centile curves for girls' weight birth to 36 months of age.

7.6 Comparisons Using ANCOVA

7.6.1 Comparing Geographical Regions

In the following analysis the aim is, for a particular measurement, sex and age group, to compare the growth trends over age in different geographical regions. These are:

i North
ii Southwest
iii Central

This means that we are looking at a large proportion of the original data used to fit the LMS models but not all of it as some individuals live in regions other than those listed above.

One approach, for a particular measurement and sex, would be to fit a different LMS model to the data in each region and then to compare the fitted models. However, we are not aware of any existing methodology to make such direct LMS model comparisons. In our proposed approach, we have taken the final LMS model fitted to all the data and used it to transform all the individual measurements into standard deviation scores.

Then, in step 1, a separate cubic regression curve was fitted, where the response ("y-variable") is the SDS score and the covariate ("x-variable") is age, to the data in each of the three regional groups. These regression lines describe how the mean SDS score of a given measurement changes with age in each region. The fit of the three cubic regression curves were then compared with the fit of three quadratic regression curves. If the difference in fits was not statistically significant, then the quadratic models were accepted and they were then compared with three linear regression curves and so on until the simplest model that might be fitted is three different constant horizontal lines. The three final regression lines can be plotted to provide a graphical description of the differences (Fig. 7.10).

If there are no differences in the three regions in how a particular measurement for a given age group and sex changes with age, then a single common regression line would be an appropriate model for all the data in the three regions. Therefore, in step 2, such a model was fitted to the data. It would be expected that it would be fairly close to the zero line but not identically zero because we have not used all the original data in this analysis as explained above. The degree of this line (cubic, quadratic, etc.) was chosen to be the same as that of the best fitting three separate ones.

The next stage is to statistically test the fit of the model involving three separate regression lines with the fit of the model based on a single common regression line. We would expect the total residual sum of the squares for the model involving three regression lines to be less than that which just involves one but we need to test whether the difference is statistically significant.

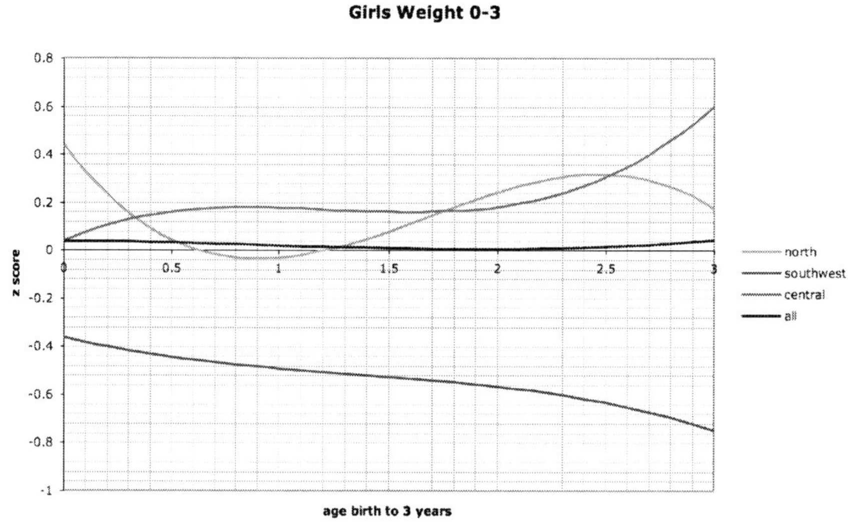

Fig. 7.10 SDS score regression models in the three geographical regions for weight vs. age; age: birth to 36 months; sex: female.

The method we have used is a standard *"F-test"* in this context, which is appropriate because the standardised data are Normally distributed. If the p-value of this test is small (less than 0.05), the conclusion would be that the single regression line is inadequate and there are significant differences between the regions in how mean SDS score of a given measurement of a given age group of a given sex changes with age (Table 7.3).

Table 7.3 Resulting p-values when testing a common regression model vs. different regression models for the three regions

age: birth to 36 months		
sex	variable	p
female	weight	$< 10^{-6}$

Finally, after finding a significant result we can then go on to use the same methodology as above but just use pairs of regions in turn to see which are significantly different from each other.

This procedure can be summarized for a given sex and measurement by the following steps:

[i] Step 1: Find the best fitting polynomials having the lowest possible common degree for each of the three regions.

[ii] Step 2: We want to answer the question "Is a common polynomial of the same degree as found in Step 1 appropriate for all three regions or do the polynomials vary with region?"
That is, for a particular measurement, sex and age group we want to test:
$H_0 : E[z \mid age] = \beta_0 + ... + \beta_q age^q$ for each region, where $q \leq 3$ is the degree of the common best fitting polynomial and $E[z|age]$ denotes the mean value of z at the given age.
vs. H_1 : The polynomial for at least two regions differ.

[iii] Step 3: After finding a significant result in Step 2 carry out pairwise comparisons between the regions.

The results of the analyses carried out in Step 2 for age birth to 3 years are given in Table 7.4. The coefficients of the polynomials in the three separate regions, as well as for all three regions together, are in Table 7.4. Those polynomials for girls' weight age birth to 3 years are plotted in Fig. 7.10. Table 7.5 details the p-values for all the pairwise comparisons between regions.

Table 7.4 Estimates of the model parameters for individual regions and all three regions together for female weight, age range birth to 36 months

sex: female					
variable	region	β_0	β_1	β_2	β_3
weight	central	0.04201	0.38775	-0.34224	0.09184
	north	0.4412	-1.2076	0.91695	-0.18133
	southwest	-0.36297	-0.20639	0.10564	-0.02648
	all	0.039995	-0.005139	-0.020645	0.007602

Table 7.5 p-values for the pairwise comparisons between the different regions using ANCOVA

sex: female, age birth to 36 months	
weight	
	p
north-central	$< 10^{-6}$
southwest-central	$< 10^{-6}$
southwest-north	$< 10^{-6}$

There are clearly significant differences between the regions for each of the measurements for both sexes in each of the age ranges.

7.6.2 Comparing Males and Females

Standard deviation scores were used to compare the growth patterns between boys and girls using very similar methodology to that described above when comparing the geographical regions. In order to make comparisons for a given measure between genders, we have used the relevant fitted girls' LMS model to standardise both girls and boys measures using the `zscores` function from Carey's [2] `lmsqreg` package.

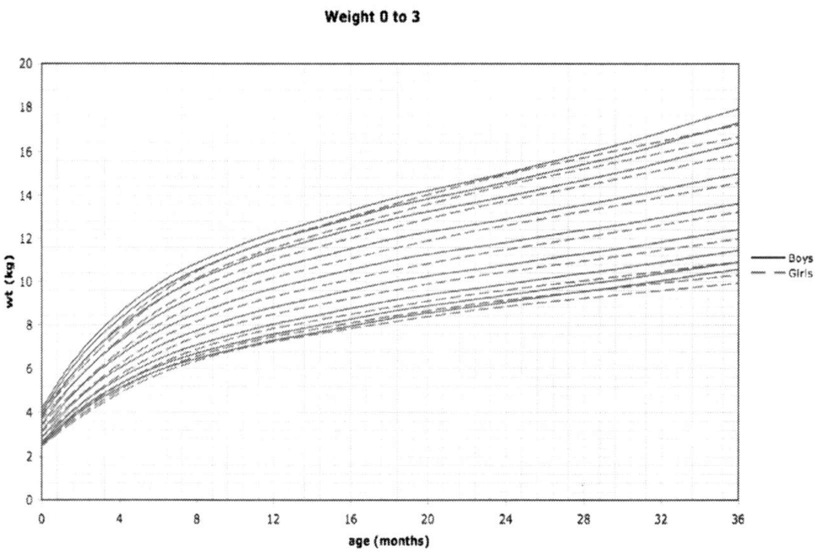

Fig. 7.11 Comparisons of the growth charts for weight measurement between males and females birth to 3 years of age.

We can then plot these standardised measures against age and construct separate regression lines for boys and girls. Considering that the data were standardised by the girls model, it is evident that the appropriate regression model for girls would be zero. However, the z scores of the boys could be explained by an appropriate polynomial regression model (up to cubic polynomial), describing the existing differences between boys and girls. If there are differences, then this will be indicated by a nonzero regression line and we can test whether the two lines are significantly different from each other using ANCOVA. We have also superimposed girls and boys centiles for a given measure on the same plot to give another graphical impression of any differences (Fig. 7.11). For children aged 0 to 3 we found significant differences for each measure and the fitted regression lines (Table 7.6) describe how the differences (measured in girls standard deviation scores) change with age (Fig. 7.12).

Table 7.6 Estimates of the model's parameters

Age birth to 36 months				
variable	β_0	β_1	β_2	β_3
length	0.21719	0.12799	-0.06722	-
head circumference	0.22312	0.75785	-0.56355	0.12686
weight	0.16774	0.70106	-0.65468	0.14854
body mass index	-	0.52468	-0.50530	0.12154

Fig. 7.12 Comparisons of growth patterns between boys and girls birth to 3 years of age.

7.7 Discussion

This study was set up by the Saudi medical authorities who required growth charts based entirely on data collected from Saudi children and adolescents rather than using a more general alternative, such as those provided by the WHO. We have seen that for girls' weight (birth to 36 months) the age-specific conditional quantile estimates we have constructed using the LMS method by and large successfully capture the main features of the data and this also proved to be true for the other growth parameters. In further work we have compared the new Saudi charts with the 2006 WHO standards and found that there are marked differences in corresponding centiles. Use of the WHO standards in Saudi Arabia would, for example, increase the prevalence of undernutrition, stunting and wasting [9].

An essential part of our procedure was to try to identify outliers to be removed from the data prior to estimating the LMS model. We used the robust regression rlm function to do this, basing our assessment on the weight attached to each observation by the procedure. We should stress that we were not using this model to make any formal inferences about the form of the conditional mean function. As seen in Sect. 7.2, this worked well with four cases being removed. If these cases were included, then there is a numerical failure in the LMS model estimation algorithm. All four deleted cases had z-scores greater than 5 in absolute value. The only other case which had an absolute z-score bigger than 5 is case 4131 with a z-score of -5.078, who can be seen listed in Fig. 7.1. This corresponds to a girl aged 0.259 year (3.11 months) who had a weight of only 3.0 kg which is a little higher than case 1235 whose weight was only 2.5 kg at a similar age and who was deleted from the data.

The evaluation of the fitted model in Sect. 7.4 indicates that the three-parameter LMS model is not always able to adequately achieve conditional normality at all ages. Under-smoothing of the parameter curves helps to remedy this problem, but at the expense of more noisy centiles. Stasinopoulos and Rigby [15] have developed the more flexible Box–Cox power exponential model, referred to as LMSP, to try to overcome this where they add an extra parameter to model kurtosis. This is implemented in the `GAMLESS R` package [16]. Wei et al. [19] advocate the use of nonparametric quantile regression methods which offer a greater degree of flexibility in their ability to model features in the conditional distribution. These are implemented in the `quantreg R` package. However, as mentioned earlier we were required to use the LMS method in this study, one of the reasons being to facilitate comparisons with existing international standards which have themselves been determined using the LMS methodology [9].

Acknowledgements We are very grateful to Professor Mohammad I. El Mouzan, MD, King Saud University, Riyadh, for providing the data. We would also like to thank the referee for helpful comments which have enabled us to improve the presentation of this paper.

References

1. Box, G.E.P., Cox, D.R.: An analysis of transformations. Journal of the Royal Statistical Society, Series B **26**, 211–252 (1964)
2. Carey, V.J.: LMSqreg: An R package for Cole–Green reference centile curves (2002). URL `http://www.biostat.harvard.edu/$\sim$carey`
3. Carey, V.J., Youg, F.H., Frenkel, L.M., McKinney, R.M.: Growth velocity assessment in pediatric aids: smoothing, penalized quantile regression and the definition of growth failure. Statistics in Medicine **23**, 509–526 (2004)
4. Carroll, R.J.: Two examples of transformations when there are possible outliers. Applied Statistics **31**, 149–152 (1982)
5. Centers for Disease Control and Prevention (CDC): Cut-offs to define outliers in the 2000 CDC Growth Charts URL `http://www.cdc.gov/nccdphp/dnpa/growthcharts/resources/BIV-cutoffs.pdf`
6. Cole, T.J.: Fitting smoothing centile curves to reference data. Journal of the Royal Statistical Society, Series A–General **151**(385–418) (1988)
7. Cole, T.J.: The lms method for constructing normalized growth standards. European Journal of Clinical Nutrition **44**(45–60) (1990)
8. Cole, T.J., Green, P.J.: Smoothing reference centile curves: the lms method and penalized likelihood. Statistics in Medicine **11**, 1305–1319 (1992)
9. El Mouzan, M.M., Foster, P.J., Al Herbish, A.S., Al Salloum, A.A., Al Omar, A.A., Qurachi, M.M., Kecojevic, T.: The implications of using the world health organization child growth standards in Saudi Arabia. Nutrition Today **44**(2), 62–70 (2009)
10. Fox, J.: An R and S-Plus companion to applied regression. SAGE Publications (2002)
11. Green, P.J.: Penalized likelihood for general semi-parametric regression models. International Statistical Review **55**(245–259) (1987)
12. Huber, P.J.: Robust estimation of a locatio parameter. Annals of Mathematical Statistics **35**, 73–101 (1964)
13. Huber, P.J.: Robust Statistics. New York: John Wiley and Sons (1981)
14. Koenker, R., Basset, G.: Regression quantiles. Econometrica **46**, 33–50 (1978)
15. Stasinopoulos, M., Rigby, B.: Smooth centile curves for skew and kurtotic data modelled using the box-cox power exponential distribution. Statistics in Medicine **23**, 3053–3076 (2004)
16. Stasinopoulos, M., Rigby, B., Akantziliotou, C.: Gamlass: An r package for generalised additive models for location, scale and shape URL `http://studweb.north.londonmet.ac.uk/\simstasinom/gamlss.html`
17. Venables, W.N., Ripley, B.D.: Mass: An r package in the standard library of venables and ripley URL `http://cran.r-project.org/src/contrib/Descriptions/VR.html`
18. Venables, W.N., Ripley, B.D.: Modern applied statistics with S, 4 edn. New York: Spriger science+busines media, Inc. (2002)
19. Wei, Y., Pere, A., Koenker, R., He, X.: Quantile regression methods for reference growth charts. Statistics in Medicine **25**, 1396–1382 (2006)

Chapter 8
Causal Mediation Analysis Using R

K. Imai, L. Keele, D. Tingley, and T. Yamamoto

Abstract Causal mediation analysis is widely used across many disciplines to investigate possible causal mechanisms. Such an analysis allows researchers to explore various causal pathways, going beyond the estimation of simple causal effects. Recently, Imai et al. (2008) [3] and Imai et al. (2009) [2] developed general algorithms to estimate causal mediation effects with the variety of data types that are often encountered in practice. The new algorithms can estimate causal mediation effects for linear and nonlinear relationships, with parametric and nonparametric models, with continuous and discrete mediators, and with various types of outcome variables. In this paper, we show how to implement these algorithms in the statistical computing language **R**. Our easy-to-use software, **mediation**, takes advantage of the object-oriented programming nature of the **R** language and allows researchers to estimate causal mediation effects in a straightforward manner. Finally, **mediation** also implements sensitivity analyses which can be used to formally assess the robustness of findings to the potential violations of the key identifying assumption. After describing the basic structure of the software, we illustrate its use with several empirical examples.

Kosuke Imai
Department of Politics, Princeton University, Princeton, NJ 08544, USA
e-mail: `kimai@princeton.edu`

Luke Keele
Department of Political Science, Ohio State University, Columbus, OH 43210, USA
e-mail: `keele.4@polisci.osu.edu`

Dustin Tingley
Department of Politics, Princeton University, Princeton, NJ 08544, USA
e-mail: `dtingley@princeton.edu`

Teppei Yamamoto
Department of Politics, Princeton University, Princeton, NJ 08544, USA
e-mail: `tyamamot@princeton.edu`

H.D. Vinod (ed.), *Advances in Social Science Research Using R*,
Lecture Notes in Statistics 196, DOI 10.1007/978-1-4419-1764-5_8,

8.1 Introduction

Causal mediation analysis is important for quantitative social science research because it allows researchers to identify possible causal mechanisms, thereby going beyond the simple estimation of causal effects. As social scientists, we are often interested in empirically testing a theoretical explanation of a particular causal phenomenon. This is the primary goal of causal mediation analysis. Thus, causal mediation analysis has a potential to overcome the common criticism of quantitative social science research that it only provides a black-box view of causality.

Recently, Imai et al. (2008) [3] and Imai et al. (2009) [2] developed general algorithms for the estimation of causal mediation effects with a wide variety of data that are often encountered in practice. The new algorithms can estimate causal mediation effects for linear and nonlinear relationships, with parametric and nonparametric models, with continuous and discrete mediators, and with various types of outcome variables. These papers [3, 2] also develop sensitivity analyses which can be used to formally assess the robustness of findings to the potential violations of the key identifying assumption. In this paper, we describe the easy-to-use software, **mediation**, which allows researchers to conduct causal mediation analysis within the statistical computing language **R** [8]. We illustrate the use of the software with some of the empirical examples presented in Imai et al. [2].

8.1.1 Installation and Updating

Before we begin, we explain how to install and update the software. First, researchers need to install **R** which is available freely at the Comprehensive R Archive Network (`http://cran.r-project.org`). Next, open **R** and then type the following at the prompt:

```
R> install.packages("mediation")
```

Once **mediation** is installed, the following command will load the package:

```
R> library("mediation")
```

Finally, to update **mediation** to its latest version, try the following command:

```
R> update.packages("mediation")
```

8.2 The Software

In this section, we give an overview of the software by describing its design and architecture. To avoid duplication, we do not provide the details of the methods that are implemented by **mediation** and the assumptions that underline them. Readers are encouraged to read Imai et al. [3, 2] for more information about the methodology implemented in **mediation**.

8.2.1 Overview

The methods implemented via **mediation** rely on the following identification result obtained under the sequential ignorability assumption of Imai et al. [3]:

$$\bar{\delta}(t) = \int\int \mathbb{E}(Y_i \mid M_i = m, T_i = t, X_i = x) \\ \left\{dF_{M_i|T_i=1,X_i=x}(m) - dF_{M_i|T_i=0,X_i=x}(m)\right\} dF_{X_i}(x), \tag{8.1}$$

$$\bar{\zeta}(t) = \int\int \{\mathbb{E}(Y_i \mid M_i = m, T_i = 1, X_i = x) \\ -\mathbb{E}(Y_i \mid M_i = m, T_i = 0, X_i = x)\}\, dF_{M_i|T_i=t,X_i=x}(m)\, dF_{X_i}(x), \tag{8.2}$$

where $\bar{\delta}(t)$ and $\bar{\zeta}(t)$ are the average causal mediation and average (natural) direct effects, respectively, and (Y_i, M_i, T_i, X_i) represents the observed outcome, mediator, treatment, and pretreatment covariates. The sequential ignorability assumption states that the observed mediator status is as if randomly assigned conditional on the randomized treatment variable and the pretreatment covariates. Causal mediation analysis under this assumption requires two statistical models: one for the mediator $f(M_i \mid T_i, X_i)$ and the other for the outcome variable $f(Y_i \mid T_i, M_i, X_i)$. (Note that we use the empirical distribution of X_i to approximate F_{X_i}.) Once these models are chosen and fitted by researchers, then **mediation** will compute the estimated causal mediation and other relevant estimates using the algorithms proposed in Imai et al. [2]. The algorithms also produce confidence intervals based on either a nonparametric bootstrap procedure (for parametric or nonparametric models) or a quasi-Bayesian Monte Carlo approximation (for parametric models).

Figure 8.1 graphically illustrates the three steps required for a mediation analysis. The first step is to fit the mediator and outcome models using, for example, regression models with the usual `lm()` or `glm()` functions. In the second step, the analyst takes the output objects from these models, which in Figure 8.1 we call `model.m` and `model.y`, and use them as inputs for the main function, `mediate()`. This function then estimates the causal mediation effects, direct effects, and total effect along with their uncertainty estimates. Finally, sensitivity analysis can be conducted via the function `medsens()` which takes the output of `mediate()` as an input. For the output of the

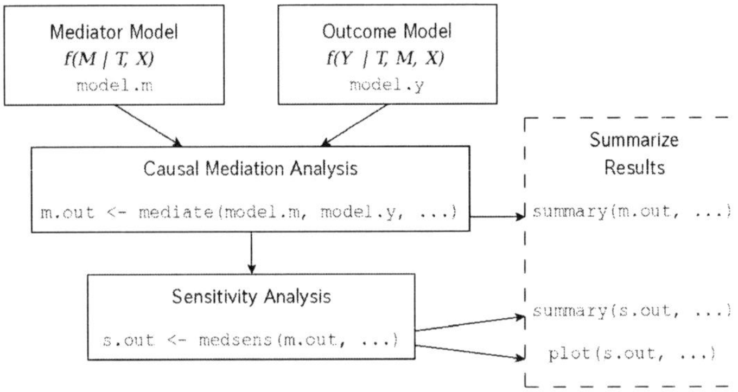

Fig. 8.1 Diagram illustrating the use of the software **mediation**. Users first fit the mediator and outcome models. Then, the function `mediate()` conducts causal mediation analysis while `medsens()` implements sensitivity analysis. The functions `summary()` and `plot()` help users interpret the results of these analyses.

`mediate()` function, a `summary()` method reports its key results in tabular form. For the output of the `medsens()` function, there are both `summary()` and `plot()` functions to display numerical and graphical summaries of the sensitivity analysis, respectively.

8.2.2 Estimation of the Causal Mediation Effects

Estimation of the causal mediation effects is based on Algorithms 1 and 2 of Imai et al. [2]. These are general algorithms in that they can be applied to any parametric (Algorithm 1 or 2) or semi/nonparametric models (Algorithm 2) for the mediator and outcome variables. Here, we briefly describe how these algorithms have been implemented in **mediation** by taking advantage of the object-oriented nature of the **R** programming language.

Algorithm 1 for Parametric Models

We begin by explaining how to implement Algorithm 1 of Imai et al. [2] for standard parametric models. First, analysts fit parametric models for the mediator and outcome variables. That is, we model the observed mediator M_i given the treatment T_i and pretreatment covariates X_i. Similarly, we model the observed outcome Y_i given the treatment, mediator, and pretreatment covariates. For example, to implement the Baron–Kenny procedure [1] in **mediation**, linear models are fitted for both the mediator and outcome models using the `lm()` command.

The model objects from these two parametric models form the inputs for the `mediate()` function. The user must also supply the names for the mediator and outcome variables along with how many simulations should be used for inference, and whether the mediator variable interacts with the treatment variable in the outcome model. Given these model objects, the estimation proceeds by simulating the model parameters based on their approximate asymptotic distribution (i.e., the multivariate normal distribution with the mean equal to the parameter estimates and the variance equal to the asymptotic variance estimate), and then computing causal mediation effects of interest for each parameter draw (e.g., using equations (8.1) and (8.2) for average causal mediation and (natural) direct effects, respectively). This method of inference can be viewed as an approximation to the Bayesian posterior distribution due to the Bernstein–von Mises Theorem [6]. The advantage of this procedure is that it is relatively computationally efficient (when compared to Algorithm 2).

We take advantage of the object-oriented nature of the **R** programming language at several steps in the function `mediate()`. For example, functions like `coef()` and `vcov()` are useful for extracting the point and uncertainty estimates from the model objects to form the multivariate normal distribution from which the parameter draws are sampled. In addition, the computation of the estimated causal mediation effects of interest requires the prediction of the mediator values under different treatment regimes as well as the prediction of the outcome values under different treatment and mediator values. This can be done by using `model.frame()` to set the treatment and/or mediator values to specific levels while keeping the values of the other variables unchanged. We then use the `model.matrix()` and matrix multiplication with the distribution of simulated parameters to compute the mediation and direct effects. The main advantage of this approach is that it is applicable to a wide range of parametric models and allows us to avoid coding a completely separate function for different models.

Algorithm 2 for Non/Semiparametric Inference

The disadvantage of Algorithm 1 is that it cannot be easily applied to non and semiparametric models. For such models, Algorithm 2, which is based on nonparametric bootstrap, can be used although it is more computationally intensive. Algorithm 2 may also be used for the usual parametric models. Specifically, in Algorithm 2, we resample the observed data with replacement. Then, for each of the bootstrapped samples, we fit both the outcome and mediator models and compute the quantities of interest. As before, the computation requires the prediction of the mediator values under different treatment regimes as well as the prediction of the outcome values under different treatment and mediator values. To take advantage of the object-oriented nature of the **R** language, Algorithm 2 relies on the `predict()` function to compute these predictions, while we again manipulate the treatment and mediator status using the `model.frame()` function. This process is repeated a large number of times and returns a bootstrap distribution of the mediation, direct, and total effects. We use the percentiles of the bootstrap distribution for confidence intervals. Thus, Algorithm 2 allows analysts to estimate mediation effects with more flexible model specifications or to estimate mediation effects for quantiles of the distribution.

8.2.3 Sensitivity Analysis

Causal mediation analysis relies on the sequential ignorability assumption that cannot be directly verified with the observed data. The assumption implies that the treatment is ignorable given the observed pretreatment confounders and that the mediator is ignorable given the observed treatment and the observed pretreatment covariates. In order to probe the plausibility of such a key identification assumption, analysts must perform a sensitivity analysis [9]. Unfortunately, it is difficult to construct a sensitivity analysis that is generally applicable to any parametric or nonparametric model. Thus, Imai et al. [3, 2] develop sensitivity analyses for commonly used parametric models, which we implement in **mediation**.

The Baron–Kenny Procedure

Imai et al. [3] develop a sensitivity analysis for the Baron–Kenny procedure and Imai et al. [2] generalize it to the linear structural equation model (LSEM) with an interaction term. This general model is given by

$$M_i = \alpha_2 + \beta_2 T_i + \xi_2^\top X_i + \varepsilon_{i2}, \tag{8.3}$$

$$Y_i = \alpha_3 + \beta_3 T_i + \gamma M_i + \kappa T_i M_i + \xi_3^\top X_i + \varepsilon_{i3}, \tag{8.4}$$

where the sensitivity parameter is the correlation between ε_{i2} and ε_{i3}, which we denote by ρ. Under sequential ignorability, ρ is equal to zero and thus the magnitude of this correlation coefficient represents the departure from the ignorability assumption (about the mediator). Note that the treatment is assumed to be ignorable as it would be the case in randomized experiments where the treatment is randomized but the mediator is not. Theorem 2 of [2] shows how the average causal mediation effects change as a function of ρ.

To obtain the confidence intervals for the sensitivity analysis, we apply the following iterative algorithm to equations (8.3) and (8.4) for a fixed value of ρ. At the tth iteration, given the current values of the coefficients, i.e., $\theta^{(t)} = (\alpha_2^{(t)}, \beta_2^{(t)}, \xi_2^{(t)}, \ldots)$, and a given error correlation ρ, we compute the variance–covariance matrix of $(\varepsilon_{i2}, \varepsilon_{i3})$, which is denoted by $\Sigma^{(t)}$. The matrix is computed by setting $\sigma_j^{(t)2} = ||\hat{\varepsilon}_j^{(t)}||^2/(n-L_j)$ and $\sigma_{23}^{(t)} = \rho\sigma_2^{(t)}\sigma_3^{(t)}$, where $\hat{\varepsilon}_j^{(t)}$

is the residual vector and L_j is the number of coefficients for the mediator model ($j = 2$) and the outcome model ($j = 3$) at the tth iteration. We then update the parameters via generalized least squares, i.e.,

$$\theta^{(t+1)} = \{V^\top ({\Sigma^{(t)}}^{-1} \otimes I_n) V\}^{-1} V^\top ({\Sigma^{(t)}}^{-1} \otimes I_n) W$$

where $V = \begin{bmatrix} \mathbf{1}\; T\; X\; \mathbf{0}\; \mathbf{0}\; \mathbf{0} & \mathbf{0} & \mathbf{0} \\ \mathbf{0}\; \mathbf{0}\; \mathbf{0}\; \mathbf{1}\; T\; M & TM & X \end{bmatrix}$, $W = \begin{bmatrix} M \\ Y \end{bmatrix}$, $T = (T_1, \ldots, T_n)^\top$, $M = (M_1, \ldots, M_n)^\top$ and $Y = (Y_1, \ldots, Y_n)^\top$ are column vectors of length n, and $X = (X_1, \ldots, X_n)^\top$ are the $(n \times K)$ matrix of observed pretreatment covariates, and $\otimes$ represents the Kronecker product. We typically use equation-by-equation least squares estimates as the starting values of θ and iterate these two steps until convergence. This is essentially an application of the iterative feasible generalized least square algorithm of the seemingly unrelated regression [12], and thus the asymptotic variance of $\hat{\theta}$ is given by $\mathrm{Var}(\hat{\theta}) = \{V^\top (\Sigma^{-1} \otimes I_n) V\}^{-1}$. Then, for a given value of ρ, the asymptotic variance of the estimated average causal mediation effects is found, for example, by the Delta method and the confidence intervals can be constructed.

The Binary Outcome Case

The sensitivity analysis for binary outcomes parallels the case when both the mediator and outcome are continuous. Here, we assume that the model for the outcome is a probit regression. Using a probit regression for the outcome allows us to assume the error terms are jointly normal with a possibly nonzero correlation ρ. Imai et al. [2] derive the average causal mediation effects as a function of ρ and a set of parameters that are identifiable due to randomization of the treatment. This lets us use ρ as a sensitivity parameter in the same way as in the Baron–Kenny procedure. For the calculation of confidence intervals, we rely on the quasi-Bayesian approach of Algorithm 1 by approximating the posterior distribution with the sampling distribution of the maximum likelihood estimates.

The Binary Mediator Case

Finally, a similar sensitivity analysis can also be conducted in a situation where the mediator variable is dichotomous and the outcome is continuous. In this case, we assume that the mediator can be modeled as a probit regression where the error term is independently and identically distributed as standard normal distribution. A linear normal regression with error variance equal to σ_3^2 is used to model the continuous outcome variable. We further assume that the two error terms jointly follow a bivariate normal distribution with mean zero and covariance $\rho\sigma_3$. Then, as in the other two cases, we use the correlation between the two error terms ρ as the sensitivity parameter. Imai et al. [2] show that under this setup, the causal mediation effects can be expressed as a function of the model parameters that can be consistently estimated given a fixed value of ρ. Uncertainty estimates are computed based on the quasi-Bayesian approach, as in the binary outcome case. The results can be graphically summarized via the `plot()` function in a manner similar to the other two cases.

Alternative Interpretations Based on R^2

The main advantage of using ρ as a sensitivity parameter is its simplicity. However, applied researchers may find it difficult to interpret the magnitude of this correlation coefficient. To overcome this limitation, Imai et al. [3] proposed alternative interpretations of ρ based on the coefficients of determination or R^2 and Imai et al. [2] extended them to the binary mediator and binary outcome cases. In that formulation, it is assumed that there exists a common unobserved pretreatment confounder in both mediator and outcome models. Applied researchers are then required to specify whether the coefficients of this unobserved confounder in the two models have the same sign or not; i.e., $\text{sgn}(\lambda_2\lambda_3) = 1$ or -1 where λ_2 and λ_3 are the coefficients in the mediator and outcome models, respectively. Once this information is provided, the average causal mediation effect can be expressed as the function of "the proportions of original variances explained by the unobserved confounder" where the original variances refer to the variances of the mediator and the outcome (or the variance of latent variable in the case of binary dependent variable). Alternatively, the average causal mediation effect can also be expressed in terms of "the proportions of the previously unexplained variances explained by the unobserved confounder" (see [1] for details). These alternative interpretations allow researchers to quantify how large the unobserved confounder must be (relative to the observed pretreatment covariates in the model) in order for the original conclusions to be reversed.

8.2.4 Current Limitations

Our software, **mediation**, is quite flexible and can handle many of the model types that researchers are likely to use in practice. Table 8.1 categorizes the types of the mediator and outcome variables and lists whether **mediation** can produce the point and uncertainty estimates of causal mediation effects. For example, while **mediation** can estimate average causal mediation effects when the mediator is ordered and the outcome is continuous, it has not yet been extended to other cases involving ordered variables. In each situation handled by **mediation**, it is possible to have an interaction term between treatment status and the mediator variable, in which case the estimated quantities of interest will be reported separately for the treatment and control groups.

Table 8.1 The types of data that can be currently handled by **mediation** for the estimation of causal mediation effects

	Outcome Variable Types		
Mediator Types	Continuous	Ordered	Binary
Continuous	Yes	No	Yes
Ordered	Yes	No	No
Binary	Yes	No	Yes

Table 8.2 The types of data that can be currently handled by **mediation** for sensitivity analysis. For continuous variables, the linear regression model is assumed. For binary variables, the probit regression model is assumed

	Outcome Variable Types		
Mediator Types	Continuous	Ordered	Binary
Continuous	Yes	No	Yes
Ordered	No	No	No
Binary	Yes	No	No

Our software provides a convenient way to probe the sensitivity of results to potential violations of the ignorability assumption for certain model types. The sensitivity analysis requires the specific derivations for each combination of models, making it difficult to develop a general sensitivity analysis method. As summarized in Table 8.2, **mediation** can handle several cases that are likely to be encountered by applied researchers. When the mediator is continuous, then sensitivity analysis can be conducted with continuous and binary outcome variables. In addition, when the mediator is binary, sensitivity analysis is available for continuous outcome variables. For sensitivity analyses that combine binary or continuous mediators and outcomes, analysts must use a probit regression model with a linear regression model. This allows for jointly normal errors in the analysis. Unlike the estimation of causal mediation effects, sensitivity analysis with treatment/mediator interactions can only be done for the continuous outcome/continuous mediator and continuous outcome/binary mediator cases. In the future, we hope to expand the range of models that are available for sensitivity analysis.

8.3 Examples

Next, we provide several examples to illustrate the use of **mediation** for the estimation of causal mediation effects and sensitivity analysis. The data used are available as part of the package so that readers can replicate the results reported below. We demonstrate the variety of models that can be used for the outcome and mediating variables.

Before presenting our examples, we load the **mediation** library and the example data set included with the library.

```
R> library("mediation")
mediation: R Package for Causal Mediation Analysis
Version: 2.0
R> data("jobs")
```

This dataset is from the Job Search Intervention Study (JOBS II) [10]. In the JOBS II field experiment, 1,801 unemployed workers received a pre-screening questionnaire and were then randomly assigned to treatment and control groups. Those in the treatment group participated in job-skills workshops. Those in the control condition received a booklet describing job-search tips. In follow-up interviews, two key outcome variables were measured: a continuous measure of depressive symptoms based on the Hopkins Symptom Checklist (`depress2`), and a binary variable representing whether the respondent had become employed (`work1`). In the JOBS II data, a continuous measure of job-search self-efficacy represents a key mediating variable (`job_seek`). In addition to the outcome and mediators, the JOBS II data also include the following list of baseline covariates that were measured prior to the administration of the treatment: pretreatment level of depression (`depress1`), education (`educ`), income, race (`nonwhite`), marital status (`marital`), age, sex, previous occupation (`occp`), and the level of economic hardship (`econ_hard`).

8.3.1 Estimation of Causal Mediation Effects

The Baron–Kenny Procedure

We start with an example when both the mediator and the outcome are continuous. In this instance, the results from either algorithm will return point estimates essentially identical to the usual Baron and Kenny procedure though the quasi-Bayesian or nonparametric bootstrap approximation is used. Using the JOBS II data, we first estimate two linear regressions for both the mediator and the outcome using the `lm()` function.

```
R> model.m <- lm(job_seek ~ treat + depress1 + econ_hard
+ sex + age + occp + marital + nonwhite + educ + income,
```

```
data = jobs)
R> model.y <- lm(depress2 ~ treat + job_seek + depress1
+ econ_hard + sex + age + occp + marital + nonwhite
+ educ + income, data = jobs)
```

These two model objects, `model.m` and `model.y`, become the arguments for the `mediate()` function. The analyst must take some care with missing values before estimating the models above. While model functions in **R** handle missing values in the data using the usual listwise deletion procedures, the functions in **mediation** assume that missing values have been removed from the data before the estimation of these two models. Thus the data for the two models must have identical observations sorted in the same order with all missing values removed. The **R** function `na.omit()` can be used to remove missing values from the data frame.

In the first call to `mediate()` below, we specify `boot = TRUE` to call the nonparametric bootstrap with 1000 resamples (`sims = 1000`). When this option is set to `FALSE` in the second call, inference proceeds via the quasi-Bayesian Monte Carlo approximation using Algorithm 1 rather than Algorithm 2. We must also specify the variable names for the treatment indicator and the mediator variable using `treat` and `mediator`, respectively.

```
R> out.1 <- mediate(model.m, model.y, sims = 1000,
boot = TRUE, treat = "treat", mediator = "job_seek")
R> out.2 <- mediate(model.m, model.y, sims = 1000,
treat = "treat", mediator = "job_seek")
```

The objects from a call to `mediate()`, i.e., `out.1` and `out.2` above, are lists which contain several different quantities from the analysis. For example, `out.1$d0` returns the point estimate for the average causal mediation effect based on Algorithm 1. The help file contains a full list of values that are contained in `mediate()` objects. The `summary()` function prints out the results of the analysis in tabular form:

```
R> summary(out.1)

 Causal Mediation Analysis

Confidence Intervals Based on Nonparametric Bootstrap

Mediation Effect:  -0.01593 95% CI  -0.031140 -0.002341
Direct Effect:  -0.03125 95% CI  -0.1045  0.0408
Total Effect:  -0.04718 95% CI  -0.11996  0.02453
Proportion of Total Effect via Mediation:
0.2882 95% CI  -2.412  3.419

R> summary(out.2)
```

```
.
Output Omitted
```

The output from the `summary()` function displays the estimates for the average causal mediation effect, direct effect, total effect, and proportion of total effect mediated. The first column displays the quantity of interest, the second column displays the point estimate, and the other columns present the 95% confidence intervals. Researchers can then easily report these point estimates and corresponding uncertainty estimates in their work. In this case, we find that job search self-efficacy mediated the effect of the treatment on depression in the negative direction. This effect, however, was small with a point estimate of $-.016$ but the 95% confidence intervals $(-.031, -.002)$ still do not contain 0.

The Baron–Kenny Procedure with the Interaction Term

Analysts can also allow the causal mediation effect to vary with treatment status. Here, the model for the outcome must be altered by including an interaction term between the treatment indicator, `treat`, and the mediator variable, `job_seek`:

```
R> model.y <- lm(depress2 ~ treat + job_seek
+ treat:job_seek + depress1 + econ_hard + sex
+ age + occp + marital + nonwhite + educ
+ income, data = jobs)
```

Users should note that under our current implementation, the interaction term must be specified in the form of `treat.name:med.name` where `treat.name` and `med.name` are the names of the treatment variable and mediator in the model, respectively. Then, a call is again made to `mediate()`, but now the option `INT = TRUE` must be specified:

```
R> out.3 <- mediate(model.m, model.y, sims = 1000,
boot = TRUE, INT = TRUE, treat = "treat", mediator =
"job_seek")
R> out.4 <- mediate(model.m, model.y, sims=1000,
INT = TRUE, treat = "treat", mediator =
"job_seek")
R> summary(out.3)

 Causal Mediation Analysis

Confidence Intervals Based on Nonparametric Bootstrap

Mediation Effect_0:  -0.02056 95% CI  -0.0425 -0.0038
Mediation Effect_1:  -0.01350 95% CI  -0.0281 -0.0023
Direct Effect_0:  -0.03318 95% CI  -0.10496  0.03592
```

```
Direct Effect_1:  -0.02611 95% CI  -0.09612  0.04454
Total Effect:  -0.04668 95% CI  -0.11594  0.02135
Proportion of Total Effect via Mediation:
0.3053 95% CI  -3.578  3.593

R> summary(out.4)
 .
 .
Output Omitted
```

Again using the `summary()` function provides a table of the results. Now estimates for the mediation and direct effects correspond to the levels of the treatment and are printed as such in the tabular summary. In this case, the mediation effect under the treatment condition, listed as `Mediation Effect_1`, is estimated to be −.014 while the mediation effect under the control condition, `Mediation Effect_0`, is −.021.

Use of Non/Semiparametric Regression

The flexibility of **mediation** becomes readily apparent when we move beyond standard linear regression models. For example, we might suspect that the mediator has a nonlinear effect on the outcome. Generalized Additive Models (GAMs) allow analysts to use splines for flexible nonlinear fits. This presents no difficulties for the `mediate()` function. We model the mediator as before, but we alter the outcome model using the `gam()` function from the `mgcv` library.

```
R> library(mgcv)
This is mgcv 1.4-1
R> model.m <- lm(job_seek ~ treat + depress1
+ econ_hard + sex + age + occp + marital
+ nonwhite + educ + income, data = jobs)
R> model.y <- gam(depress2 ~ treat + s(job_seek,
bs = "cr")  + depress1 + econ_hard + sex + age
+ occp + marital + nonwhite + educ + income,
data = jobs)
```

In this case we fit a Generalized Additive Model for the outcome variable, and allow the effect of the `job_seek` variable to be nonlinear and determined by the data. This is done by using the `s()` notation which allows the fit between the mediator and the outcome to be modeled with a spline. Using the spline for the fit allows the estimate for the mediator on the outcome to be a series of piecewise polynomial regression fits. This semiparametric regression model is a more general version of nonparametric regression models such as lowess. The model above allows the estimate to vary across the range of the predictor variable. Here, we specify the model with a cubic basis function (`bs = "cr"`) for the smoothing spline and leave the smoothing selection to be done at the program defaults which is generalized cross-validation. Fully understanding how to fit such models is beyond the scope here. Interested readers should consult Wood 2006 [11] for full technical details and Keele 2008 [5] provides coverage of these models from a social science perspective.

The call to `mediate()` with a `gam()` fit remains unchanged except that when the outcome model is a semiparametric regression only the nonparametric bootstrap is valid for calculating uncertainty estimates, i.e., `boot = TRUE`.

```
R> out.5 <- mediate(model.m, model.y, sims = 1000,
boot = TRUE, treat = "treat", mediator = "job_seek")

R> summary(out.5)
.
.
Output Omitted
```

The model for the mediator can also be modeled with the `gam()` function as well. The `gam()` function also allows analysts to include interactions; thus analysts can still allow the mediation effects to vary with treatment status. This simply requires altering the model specification by using the `by` option in the `gam()` function and using two separate indicator variables for treatment status. To fit this model we need one variable that indicates whether the observation was in the treatment group and a second variable that indicates whether the observation was in the control group. To allow the mediation effect to vary with treatment status, the call to `gam()` takes the following form:

```
R> model.y <- gam(depress2 ~ treat + s(job_seek, by = treat)
+ s(job_seek, by = control) + depress1 + econ_hard + sex
+ age + occp + marital + nonwhite + educ + income,
data = jobs)
```

In this case, we must also alter the options in the `mediate()` function by specifying `INT = TRUE` and provide the variable name for the control group indicator using the `control` option.

```
R> out.6 <- mediate(model.m, model.y, sims = 1000,
boot = TRUE, INT = TRUE, treat = "treat",
mediator = "job_seek", control = "control")

R> summary(out.6)

 Causal Mediation Analysis

Confidence Intervals Based on Nonparametric Bootstrap

Mediation Effect_0:  -0.02328 95% CI  -0.059138  0.006138
```

```
Mediation Effect_1:  -0.01622 95% CI  -0.041565  0.004363
Direct Effect_0:  -0.01408 95% CI  -0.09369  0.05672
Direct Effect_1:  -0.007025 95% CI  -0.08481  0.06114
Total Effect:  -0.0303 95% CI  -0.13065  0.04744
Proportion of Total Effect via Mediation:
0.3395% CI  -8.514  4.391
```

As the reader can see, despite the fact that the mediator was specified as a nonparametric function, one still receives point estimates and confidence intervals for the mediation effect across each treatment level. In the table, `Mediation Effect_0` and `Direct Effect_0` are the mediation and direct effects respectively under the control condition, while `Mediation Effect_1` and `Direct Effect_1` are the mediation and direct effects under treatment.

Quantile Causal Mediation Effects

Researchers might also be interested in modeling mediation effects for quantiles of the outcome. Quantile regression allows for a convenient way to model the quantiles of the outcome distribution while adjusting for a variety of covariates [7]. For example, a researcher might be interested in the 0.5 quantile (i.e., median) of the distribution. This also presents no difficulties for the `mediate()` function. Again for these models, uncertainty estimates are calculated using the nonparametric bootstrap. To use quantile regression, we load the `quantreg` library and model the median of the outcome, though other quantiles are also permissible. Analysts can also relax the no-interaction assumption for the quantile regression models as well. Below we estimate the mediator with a standard linear regression, while for the outcome we use `rq()` to model the median.

```
R> library(quantreg)
Loading required package: SparseM
Package SparseM (0.78) loaded.
To cite, see citation("SparseM")
Package quantreg (4.26) loaded.
To cite, see citation("quantreg")

R> model.m <- lm(job_seek ~ treat + depress1 + econ_hard
+ sex + age + occp + marital + nonwhite + educ + income,
data = jobs)
R> model.y <- rq(depress2 ~ treat + job_seek + depress1
+ econ_hard + sex + age + occp + marital + nonwhite
+ educ + income, tau= 0.5, data = jobs)
R> out.7 <- mediate(model.m, model.y, sims = 1000,
boot = TRUE, treat = "treat", M = "job_seek")

R> summary(out.7)
```

```
 Causal Mediation Analysis

Confidence Intervals Based on Nonparametric Bootstrap

Mediation Effect:  -0.01470 95% CI  -0.027235 -0.001534
Direct Effect:  -0.02489 95% CI  -0.09637  0.04309
Total Effect:  -0.03959 95% CI  -0.11523  0.02857
Proportion of Total Effect via Mediation:
0.3337 95% CI  -3.069  1.902
```

where the `summary()` command gives the estimated median causal mediation effect along with the estimates for the other quantities of interest.

It is also possible to estimate mediation effects for quantiles of the outcome other than the median. This is done simply by specifying a different outcome quantile in the quantile regression model. For example, if the 10th percentile of the outcome were of interest, then the user can change the `tau` option,

```
R> model.y <- rq(depress2 ~ treat + job_seek + depress1
+ econ_hard + sex + age + occp + marital + nonwhite
+ educ + income, tau = 0.1, data = jobs)
```

Furthermore, it is straightforward to loop over a set of quantiles and graph the mediation effects for a range of quantiles, as done in [2].

Discrete Mediator and Outcome Data

Often analysts use measures for the mediator and outcome that are discrete. For standard methods, this has presented a number of complications requiring individually tailored techniques. The **mediation** software, however, can handle a number of different discrete data types using the general algorithms developed in Imai et al. [2]. For example, one outcome of interest in the JOBS II study is a binary indicator (`work1`) for whether the subject became employed after the training sessions. To estimate the mediation effect, we simply use a probit regression instead of a linear regression for the outcome and then call `mediate()` as before:

```
R> model.m <- lm(job_seek ~ treat + depress1 + econ_hard
+ sex + age + occp + marital + nonwhite + educ + income,
data = jobs)
R> model.y <- glm(work1 ~ treat + job_seek + depress1
+ econ_hard + sex + age + occp + marital + nonwhite + educ
+ income, family = binomial(link = "probit"), data = jobs)
R> out.8 <- mediate(model.m, model.y, sims = 1000,
boot = TRUE, treat = "treat", mediator = "job_seek")
R> out.9 <- mediate(model.m, model.y, sims = 1000,
treat = "treat", mediator = "job_seek")
```

```
R> summary(out.8)
.
.
Output Omitted

R> summary(out.9)

 Causal Mediation Analysis

Quasi-Bayesian Confidence Intervals

Mediation Effect:  0.003780 95% CI  -0.0005248  0.0109583
Direct Effect:  0.05573 95% CI  -0.007416  0.119900
Total Effect:  0.05951 95% CI  -0.004037  0.123071
Proportion of Total Effect via Mediation:
0.05804 95% CI  -0.2405  0.4498
```

In the table printed by the `summary()` function, the estimated average causal mediation effect along with the quasi-Bayesian confidence interval are printed on the first line followed by the direct and total effects, and the proportion of the total effect due to mediation. It is also possible to use a logit model for the outcome instead of a probit model. However, we recommend the use of a probit model because our implementation of the sensitivity analysis below requires a probit model for analytical tractability.

The mediator can also be binary or an ordered measure as well. This simply requires modeling the mediator with either a probit or ordered probit model. For demonstration purposes, the `jobs` data contains two variables, `job_dich` and `job_disc`, which are recoded versions of `job_seek`. The first measure is simply the continuous scale divided at the median into a binary variable. The second measure, `job_disc`, recodes the continuous scale into a discrete four-point scale. We emphasize that this is for demonstration purposes only, and analysts in general should not recode continuous measures into discrete measures. Estimating mediation effects with a binary mediator is quite similar to the case above with a binary outcome. We simply now use a probit model for the mediator and a linear regression for the outcome:

```
R> model.m <- glm(job_dich ~ treat + depress1 + econ_hard
+ sex + age + occp + marital + nonwhite + educ + income,
data = job, family = binomial(link = "probit"))
R> model.y <- lm(depress2 ~ treat + job_dich + treat:job_dich
+ depress1 + econ_hard + sex + age + occp + marital
+ nonwhite + educ + income, data = jobs)
```

In this example we allow the effect of the mediator to vary with treatment status. The user now calls `mediate()` and can use either the quasi-Bayesian approximation or nonparametric bootstrap.

```
R> out.10 <- mediate(model.m, model.y, sims = 1000,
boot=TRUE, treat="treat", mediator="job_dich", INT=TRUE)
R> out.11 <- mediate(model.m, model.y, sims = 1000,
treat = "treat", mediator = "job_dich", INT = TRUE)
R> summary(out.10)
.
.
Output Omitted
R> summary(out.11)
 Causal Mediation Analysis

Quasi-Bayesian Confidence Intervals

Mediation Effect_0:  -0.01809 95% CI  -0.035290 -0.005589
Mediation Effect_1:  -0.01968 95% CI  -0.034518 -0.007263
Direct Effect_0:  -0.02849 95% CI  -0.1008  0.0393
Direct Effect_1:  -0.03009 95% CI  -0.10111  0.03791
Total Effect:  -0.04817 95% CI  -0.11962  0.01729
Proportion of Total Effect via Mediation:
0.3431 95% CI  -3.330  3.756
```

In the table, we see that `Mediation Effect_0` is the mediation effect under the control condition, while `Mediation Effect_1` is the mediation effect under the treatment condition. The same notation applies to the direct effects. As the reader can see, the output also indicates which algorithm is used for the 95% confidence intervals.

When the mediator is an ordered variable, we switch to an ordered probit model for the mediator. In **R**, the `polr()` function in the `MASS` library provides this functionality. The `MASS` library is automatically loaded with **mediation** so the `polr()` function is readily available to users. Thus, we fit the outcome and mediator models below:

```
R> model.m <- polr(job_disc ~ treat + depress1 + econ_hard
+ sex + age + occp + marital + nonwhite + educ + income,
data = jobs, method = "probit", Hess = TRUE)
R> model.y <- lm(depress2 ~ treat + job_disc + depress1
+ econ_hard + sex + age + occp + marital + nonwhite
+ educ + income, data = jobs)
```

The reader should note that in the call to `polr()` the `Hess = TRUE` needs to be specified to use the quasi-Bayesian approximation in the `mediate()` function. Once we have estimated these two models, analysis proceeds as before:

```
R> out.12 <- mediate(model.m, model.y, sims = 1000,
boot = TRUE, treat = "treat", mediator = "job_disc")
R> out.13 <- mediate(model.m, model.y, sims = 1000,
```

```
treat = "treat", mediator = "job_disc")

R> summary(out.12)
 .
 .
Output Omitted
R> summary(out.13)
 .
 .
Output Omitted
```

Again, for any of these data types, analysts can relax the no-interaction assumption as before by including the interaction between treatment and the mediator variable in the outcome model and using the `INT = TRUE` option.

8.3.2 Sensitivity Analysis

Once analysts have estimated mediation effects, they should always explore how robust their finding is to the ignorability assumption. The `medsens()` function allows analysts to conduct sensitivity analyses for mediation effects. Next, we provide a demonstration of the functionality for the sensitivity analysis. Currently, **mediation** can conduct sensitivity analyses for the continuous–continuous case, the binary–continuous case, and the continuous–binary case.

The Baron–Kenny Procedure

As before, one must first fit models for the mediator and outcome and then pass these model objects through the mediate function:

```
R> model.m <- lm(job_seek ~ treat + depress1 + econ_hard
+ sex + age + occp + marital + nonwhite + educ + income,
data = jobs)
R> model.y <- lm(depress2 ~ treat + job_seek + depress1
+ econ_hard + sex + age + occp
+ marital + nonwhite + educ + income, data = jobs)
R> med.cont <- mediate(model.m, model.y, sims=1000,
treat = "treat", mediator = "job_seek")
```

Once the analyst estimates the mediation effects, the output from the `mediate()` function becomes the argument for `medsens()`, which is the workhorse function. The `medsens()` function recognizes the options specified in the `mediate()` function and thus there is no need to specify the `treat`, `mediator`, or `INT` options.

```
R> sens.cont <- medsens(med.cont, rho.by = 0.05)
```

The `rho.by` option specifies how finely incremented the parameter ρ is for the sensitivity analysis. Using a coarser grid for ρ speeds up estimation considerably, but this comes at the cost of estimating the robustness of the original conclusion only imprecisely.

After running the sensitivity analysis via `medsens()`, the `summary()` function can be used to produce a table with the values of ρ for which the confidence interval contains zero. This allows the analyst to immediately see the approximate range of ρ where the sign of the causal mediation effect is indeterminate. The second section of the table contains the value of ρ for which the mediation effect is exactly zero, which in this application is -0.19. The table also presents coefficients of determination that correspond to the critical value of ρ where the mediation effect is zero. First, $R^{*2}_{M}R^{*2}_{Y}$ is the product of coefficients of determination which represents the proportion of the *previously unexplained* variance in the mediator and outcome variables that is explained by an unobservable pretreatment unconfounder. An alternative formulation is in terms of the proportion of the *original* variance explained by an unobserved confounder, which we denote as $\widetilde{R}^2_M\widetilde{R}^2_Y$.

```
R> summary(sens.cont)

Mediation Sensitivity Analysis

Sensitivity Region
      Rho Med. Eff. 95% CI    95% CI  R^2_M*R^2_Y* R^2_M~R^2_Y
                     Lower     Upper
[1,] -0.25   0.0056  -0.0008   0.0120    0.0625          0.0403
[2,] -0.20   0.0012  -0.0035   0.0058    0.0400          0.0258
[3,] -0.15  -0.0032  -0.0084   0.0020    0.0225          0.0145
[4,] -0.10  -0.0074  -0.0150   0.0001    0.0100          0.0064

Rho at which ACME = 0: -0.1867

R^2_M*R^2_Y* at which ACME = 0: 0.0349

R^2_M~R^2_Y~ at which ACME = 0: 0.0225
```

The table above presents the estimated mediation effect along with its confidence interval for each value of ρ. The reader can verify that when ρ is equal to zero, the reported mediation effect matches the estimate produced by the `mediate()` function. For other values of ρ, the mediation effect is calculated under different levels of unobserved confounding.

The information from the sensitivity analysis can also be summarized graphically using the `plot()` function. First, passing the `medsens` object to `plot()` and specifying the `sens.par` option to `"rho"`, i.e.,

```
R> plot(sens.cont, sens.par = "rho")
```

produces the left-hand side of Figure 8.2. In the plot, the dashed horizontal line represents the estimated mediation effect under the sequential ignorability assumption, and the solid line represents the mediation effect under various values of ρ. The gray region represents the 95% confidence bands.

Similarly, we can also plot the sensitivity analysis in terms of the coefficients of determination as discussed above. Here we specify `sens.par` option to `"R2"`. We also need to specify two additional pieces of information. First, `r.type` option tells the plot function whether to plot $R^{*2}_M R^{*2}_Y$ or $\widetilde{R}^2_M \widetilde{R}^2_Y$. To plot the former `r.type` is set to `1` and to plot the latter `r.type` is set to `2`. Finally, the `sign.prod` option specifies the sign of the product of the coefficients of the unobserved confounder in the mediator and outcome models. This product indicates whether the unobserved confounder affects both mediator and outcome variables in the same direction (`1`) or different directions (`-1`), thereby reflecting the analyst's expectation about the nature of confounding.

For example, the following command produces the plot representing the sensitivity of estimates with respect to the proportion of the original variances explained by the unobserved confounder when the confounder is hypothesized to affect the mediator and outcome variables in opposite directions.

```
R> plot(sens.cont, sens.par = "R2", r.type = 2,
sign.prod = -1)
```

The resulting plot is shown on the right-hand side of Figure 8.2. Each contour line represents the mediation effect for the corresponding values of $\widetilde{R}^2_M$ and $\widetilde{R}^2_Y$. For example, the 0 contour line corresponds to values of the product $\widetilde{R}^2_M \widetilde{R}^2_Y$ such that the average causal mediation effect is 0. As reported in the table, even a small proportion of original variance unexplained by the confounder, .02%, produces mediation effects of 0. Accordingly, the right-hand side of Figure 8.2 shows how increases in $\widetilde{R}^2_M \widetilde{R}^2_Y$ (moving from the lower left to upper right) produce *positive* mediation effects.

For both types of sensitivity plots, the user can specify additional options available in the plot function such as alternative title (`main`) and axis labels (`xlab`, `ylab`) or manipulate common graphical options (e.g., `xlim`).

Binary Outcome

The `medsens()` function also extends to analyses where the mediator is binary and the outcome is continuous, as well as when the mediator is continuous and the outcome is binary. If either variable is binary, `medsens()` takes an additional argument. For example, recall the binary outcome model estimated earlier:

```
R> model.y <- glm(work1 ~ treat + job_seek + depress1
+ econ_hard + sex + age + occp + marital + nonwhite
```

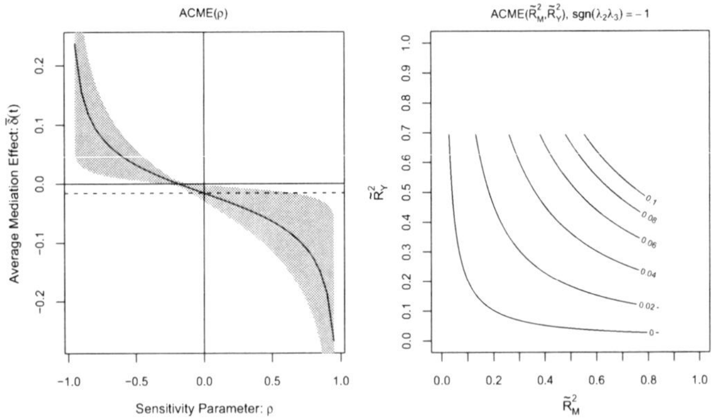

Fig. 8.2 Sensitivity analysis with continuous outcome and mediator.

```
+ educ  + income, family = binomial(link = "probit"),
data = jobs)
R> med.bout <- mediate(model.m, model.y, sims = 1000,
treat = "treat", mediator = "job_seek")
```

The call to `medsens()` works as before, with the output from the `mediate()` function passed through `medsens()`.

```
R> sens.bout <- medsens(med.bout, rho.by = 0.05,
sims = 1000)
```

The `sims` option provides control over the number of draws in the parametric bootstrap procedure which is used to compute confidence bands. When either the mediator or outcome is binary, the exact values of sensitivity parameters where the mediation effects are zero cannot be analytically obtained as in the fully continuous case (see [3] Section 4). Thus, this information is reported based on the signs of the estimated mediation effects under various values of ρ and corresponding coefficients of determination. The usage of the `summary()` function, however, remains identical to the fully continuous case in that the output table contains the estimated mediation effects and the corresponding values of ρ for which the confidence region contains zero.

As in the case with continuous mediator and outcome variables, we can plot the results of the sensitivity analysis. The following code produces Figure 8.3.

```
R> plot(sens.bout, sens.par = "rho")
R> plot(sens.bout, sens.par = "R2", r.type = 2,
sign.prod = 1)
```

On the left-hand side we plot the average causal mediation effects in terms of ρ, while we use $\widetilde{R}_M^2$ and $\widetilde{R}_Y^2$ on the right-hand side. In the ρ plot, the dashed line represents the estimated mediation effect under sequential ignorability,

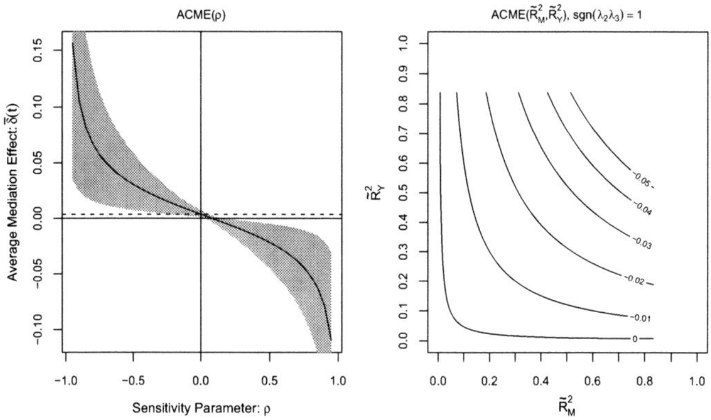

Fig. 8.3 Sensitivity analysis with continuous outcome and binary mediator.

and the solid line represents the mediation effect under various values of ρ. The gray region represents the 95% confidence bands. In the $\widetilde{R}^2$ plot the average causal mediation effect is plotted against various values of $\widetilde{R}^2_M$ and $\widetilde{R}^2_Y$ and is interpreted in the same way as above.

When the outcome is binary, the proportion of the total effect due to mediation can also be calculated as a function of the sensitivity parameter ρ. The `pr.plot` option in the plot command (in conjunction with the `sens.par = "rho"` option) allows users to plot a summary of the sensitivity analysis for the proportion mediated. For example, the following call would provide a plot of this quantity:

```
R> plot(sens.bout, sens.par = "rho", pr.plot = TRUE)
```

Binary Mediator

The final form of sensitivity analysis deals with the case where the outcome variable is continuous but the mediator is binary. For the purpose of illustration, we simply dichotomize the `job_seek` variable to produce a binary measure `job_dich`. We fit a probit model for the mediator and linear regression for the outcome variable.

```
R> model.m <- glm(job_dich ~ treat + depress1
+ econ_hard + sex + age + occp + marital + nonwhite
+ educ + income, data = jobs,
family = binomial(link = "probit"))
R> model.y <- lm(depress2 ~ treat + job_dich+ depress1
+ econ_hard + sex + age + occp
+ marital + nonwhite + educ + income, data = jobs)
R> med.bmed <- mediate(model.m, model.y, sims = 1000,
treat = "treat", mediator = "job_dich")
```

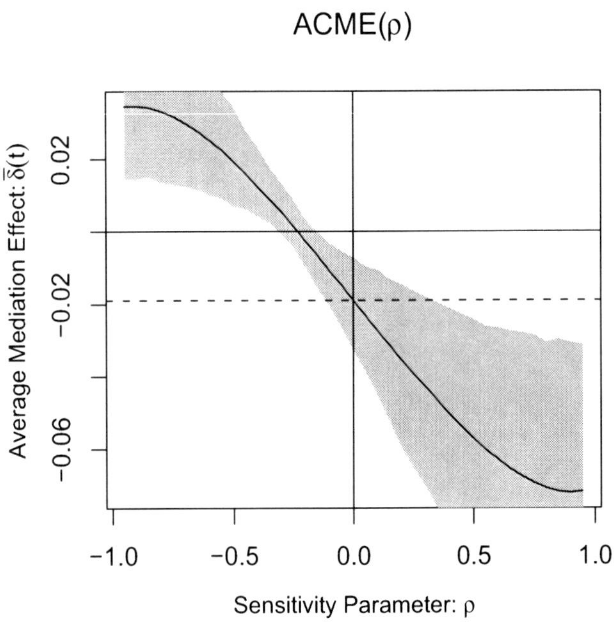

Fig. 8.4 Sensitivity analysis with continuous outcome and binary mediator.

```
R> sens.bmed <- medsens(med.bmed, rho.by = 0.05,
sims = 1000)
```

Again we can pass the output of the `medsens()` function through the `plot()` function:

```
R> plot(sens.bmed, sens.par = "rho")
```

producing Figure 8.4. The plot is interpreted in the same way as the above cases. The user also has the option to plot sensitivity results in terms of the coefficients of determination just as in the case with continuous outcome and mediator variables.

When the mediator variable is binary, the plotted values of the mediation effect and their confidence bands may not be perfectly smooth curves due to simulation errors. This is especially likely when the number of simulations (`sims`) is set to a small value. In such situations, the user can choose to set the `smooth.effect` and `smooth.ci` options to `TRUE` in the `plot()` function so that the corresponding values become smoothed out via a lowess smoother before being plotted. Although this option often makes the produced graph look nicer, the user should be cautious as the adjustment could affect one's substantive conclusions in a significant way. A recommended alternative is to increase the number of simulations.

8.4 Concluding Remarks

Causal mediation analysis is a key tool for social scientific research. In this paper, we describe our easy-to-use software for causal mediation analysis, **mediation**, that implements the new methods and algorithms introduced by Imai et al. 2008 [3] and Imai et al. 2009 [2]. The software provides a flexible, unified approach to causal mediation analysis in various situations encountered by applied researchers. The object-oriented nature of the **R** programming made it possible for us to implement these algorithms in a fairly general way. In addition to the estimation of causal mediation effects, **mediation** implements formal sensitivity analyses so that researchers can assess the robustness of their findings to the potential violations of the key identifying assumption. This is an important contribution for at least two reasons. First, even in experiments with randomize treatments, causal mediation analysis requires an additional assumption that is not directly testable from the observed data. Thus, researchers must evaluate the consequences of potential violations of the assumption via sensitivity analysis. Alternatively, researchers might use other experimental designs though this entails making other assumptions [4]. Second, the accumulation of such sensitivity analyses is essential for interpreting the relative degree of robustness across different studies. Thus, the development of easy-to-use software, such as **mediation**, facilitates causal mediation analysis in applied social science research in several critical directions.

8.5 Notes and Acknowledgment

The most recent version (along with all previous versions) of the **R** package, **mediation**, is available for download at the Comprehensive R Archive Network (`http://cran.r-project.org/web/packages/mediation`). This article is based on version 2.1 of **mediation**. Financial support from the National Science Foundation (SES-0849715 and SES-0918968) is acknowledged.

References

1. Baron, R., Kenny, D.: The moderator-mediator variable distinction in social psychological research: Conceptual, strategic, and statistical considerations. Journal

of Personality and Social Psychology **51**(4), 1173–1182 (1986)
2. Imai, K., Keele, L., Tingley, D.: A general approach to causal mediation analysis. Tech. rep., Department of Politics, Princeton University (2009). Available at `http://imai.princeton.edu/research/BaronKenny.html`
3. Imai, K., Keele, L., Yamamoto, T.: Identification, inference and sensitivity analysis for causal mediation effects. Tech. rep., Department of Politics, Princeton University (2008). Available at `http://imai.princeton.edu/research/mediation.html`
4. Imai, K., Tingley, D., Yamamoto, T.: Experimental identification of causal mechanisms. Tech. rep., Department of Politics, Princeton University (2009). Available at `http://imai.princeton.edu/research/Design.html`
5. Keele, L.: Semiparametric Regression for the Social Sciences. Wiley and Sons, Chichester, UK (2008)
6. King, G., Tomz, M., Wittenberg, J.: Making the most of statistical analyses: Improving interpretation and presentation. American Journal of Political Science **44**, 341–355 (2000)
7. Koenker, R.: Quantile Regression. Cambridge University Press, Cambridge (2008)
8. R Development Core Team: R: A Language and Environment for Statistical Computing. R Foundation for Statistical Computing, Vienna, Austria (2009). URL `http://www.R-project.org`. ISBN 3-900051-07-0
9. Rosenbaum, P.R.: Observational Studies, 2nd edn. Springer-Verlag, New York (2002)
10. Vinokur, A., Schul, Y.: Mastery and inoculation against setbacks as active ingredients in the jobs intervention for the unemployed. Journal of Consulting and Clinical Psychology **65**(5), 867–877 (1997)
11. Wood, S.: Generalized Additive Models: An Introduction With R. Chapman & Hall/CRC, Boca Raton (2006)
12. Zellner, A.: An efficient method of estimating seemingly unrelated regressions and tests for aggregation bias. Journal of the American Statistical Association **57**, 348–368 (1962)

Chapter 9
Statistical Validation of Functional Form in Multiple Regression Using R

Harry Haupt, Joachim Schnurbus, and Rolf Tschernig

Abstract In applied statistical research the practitioner frequently faces the problem that there is neither clear guidance from grounds of theoretical reasoning nor empirical (meta) evidence on the choice of functional form of a tentative regression model. Thus, parametric modeling resulting in a parametric benchmark model may easily miss important features of the data. Using recently advanced nonparametric regression methods we illustrate two powerful techniques to validate a parametric benchmark model. We discuss an empirical example using a well-known data set and provide R code snippets for the implementation of simulations and examples.

9.1 Model Validation

Consider a typical situation in applied multiple regression analysis where we wish to explain a continuous scalar response variable y by a set of say K potential explanatory variables collected in the vector $\mathbf{x}$, where $\mathbf{x}$ may contain both continuous and (ordered and unordered) categorical variables, for a given data sample of size n. Let us assume in addition that there are no restrictions from economic, ecologic, psychologic, etc. theory about the

Harry Haupt
Centre for Statistics, Department of Economics and Business Administration, Bielefeld University, 33501 Bielefeld, Germany, e-mail: hhaupt@wiwi.uni-bielefeld.de

Joachim Schnurbus
Institute of Economics and Econometrics, University of Regensburg, 93053 Regensburg, Germany, e-mail: joachim.schnurbus@wiwi.uni-regensburg.de

Rolf Tschernig
Institute of Economics and Econometrics, University of Regensburg, 93053 Regensburg, Germany, e-mail: rolf.tschernig@wiwi.uni-regensburg.de

H.D. Vinod (ed.), *Advances in Social Science Research Using R*,
Lecture Notes in Statistics 196, DOI 10.1007/978-1-4419-1764-5_9,

functional relationship between y and $\mathbf{x}$, and—if there are—any unknown model parameters, say $\boldsymbol{\beta}$.

It is common practice to consider a class of parametric models such as classical multiple linear regression models (or other members of the Box–Cox family of transformations) and, following some model selection procedure(s), to choose and estimate a specific parsimonious parametric model. In the sequel we refer to this model as the parametric benchmark model. Before using the benchmark model it should be validated or, more precisely, subjected to diagnostic tests, among them misspecification tests. Such tests should allow for a wide range of models under the alternative. This can be accomplished, for example, by considering nonparametric tests. Of course, this requires that the estimation of nonparametric models is appropriate and that the resulting statistical inference is informative. This implies that given the number of continuous covariates the number of observations is sufficiently large. Recently, Hsiao, Li, and Racine (see [3]) suggested a test for parametric misspecification that is implemented in the `np` package for R of Hayfield and Racine (see [2]). In comparison to other nonparametric misspecification tests it uses information from discrete covariates more efficiently.

As an alternative to misspecification tests one may compare competing models with respect to their predictive ability. A common quantity for measuring predictive quality is the mean (weighted) integrated squared error of prediction (compare e.g. Sect. 2.2.1 in [8]). Since this quantity is unobservable, it has to be estimated. In a cross-sectional context this can be done by repeatedly (B times) drawing shuffled random subsamples from the data set. For each replication, one part of the data is used for estimation, the other part for prediction. From the latter data we compute the average squared error of prediction (ASEP) defined below in (9.4). The ASEP provides an estimate of the (weighted) integrated squared error or prediction. By averaging over all replications one obtains an estimate of the *mean* (weighted) integrated squared error or prediction. This procedure is applied to both the benchmark and the alternative model and one may then test for the equality of the prediction measures. Further, one may compare the ASEP of both models for each replication and compute the percentage in which the benchmark parametric model outperforms the alternative model with respect to ASEP. That is, the ASEP of the benchmark model is smaller than that of the alternative model in $\vartheta \cdot 100$ percent of the B replications. Some computational details can be found in Sect. 9.2.

In many cases one may also compare some model selection criteria. All three approaches mentioned for validation contain some feature to punish overfitting that would result by choosing models on the basis of simple goodness-of-fit measures such as R^2 or some Pseudo-R^2 such as (9.3) below. It is worth noting that the choice of relevant criteria to choose between models, say a parametric benchmark model $\mathbb{M}_{\mathsf{p}}$, and a semi- or nonparametric model, say $\mathbb{M}_{\mathsf{np}}$, naturally depends on the objective of the empirical analysis. For example, the objective may be the analysis of marginal effects or prediction or both. However, it is widespread practice to use model selection procedures and diagnostics independently of the specific modeling goal. Although this is clearly not optimal, we do not discuss this issue further.

In this paper we illustrate the benefits and drawbacks of validation techniques based on misspecification tests and prediction simulations drawing heavily on modern nonparametric regression techniques implemented in R (version 2.9.1). We use the `np` package (version 0.30-3) of Hayfield and Racine (see [2]) for computing flexible nonparametric kernel regression models and conducting the nonparametric misspecification test of [3]. In addition, we show how the `relax` package (version 1.2.1) of Wolf (see [11]) can be used for interactive graphical analysis in R. Clever graphical analysis can help to identify data points that cause a rejection of the benchmark parametric model and to specify an alternative parametric model. We also indicate how the presented code-snippets can be generalized. The empirical illustration follows the seminal work of Hamermesh and Biddle (see [1]) on the impact of looks on earning.

The remainder of the paper is organized as follows. In Sect. 9.2 we briefly describe a general approach for the validation of models with continuous and discrete covariates. Sect. 9.3 considers `relax` and its use for different steps of the validation process. The empirical illustration is contained in Sect. 9.4.

9.2 Nonparametric Methods for Model Validation

In this section we briefly introduce nonparametric regression and misspecification testing. We are interested in modeling the conditional mean

$$E[y|\mathbf{x},\mathbf{z}] = f(\mathbf{x},\mathbf{z}), \tag{9.1}$$

where we explicitly allow for discrete and continuous covariates, collected in vectors $\mathbf{z}$ and $\mathbf{x}$, respectively. The conditional mean (9.1) may be modeled parametrically

$$E[y|\mathbf{x},\mathbf{z}] = f(\mathbf{x},\mathbf{z},\beta) \tag{9.2}$$

or nonparametrically. In the latter case, we use the mixed kernel regression approach of Li and Racine (see [7], [8], and [9]) where discrete and continuous covariates are smoothed simultaneously using specific weighting functions (kernels) for ordered and unordered discrete as well as continuous variables. The minimization calculus for a local linear regression at position $(\mathbf{x}_0,\mathbf{z}_0)$ is given by

$$\min_{\tilde{b}_0(\mathbf{x}_0,\mathbf{z}_0),\tilde{\mathbf{b}}_1(\mathbf{x}_0,\mathbf{z}_0)} \sum_{i=1}^{n} \left(y_i - \tilde{b}_0(\mathbf{x}_0,\mathbf{z}_0) - \tilde{\mathbf{b}}_1(\mathbf{x}_0,\mathbf{z}_0)^{\top} \cdot (\mathbf{x}_i - \mathbf{x}_0)\right)^2 \cdot W(\mathbf{x}_0,\mathbf{x}_i,\mathbf{z}_0,\mathbf{z}_i,\mathbf{h}),$$

where $\widehat{b}_0(\mathbf{x}_0,\mathbf{z}_0)$ estimates the conditional mean (9.1) at position $(\mathbf{x}_0,\mathbf{z}_0)$, while $\widehat{\mathbf{b}}_1(\mathbf{x}_0,\mathbf{z}_0)$ captures the partial effects of all continuous covariates. The observations in the neighborhood of $(\mathbf{x}_0,\mathbf{z}_0)$ are used to calculate values $\widehat{b}_0(\mathbf{x}_0,\mathbf{z}_0)$ and $\widehat{\mathbf{b}}_1(\mathbf{x}_0,\mathbf{z}_0)$, where the neighborhood is determined by the weighting function $W(.)$. The latter is denoted as generalized product kernel, as it depends on $\mathbf{x}_0,\mathbf{x}_i,\mathbf{z}_0,\mathbf{z}_i$, and the vector of smoothing parameters $\mathbf{h}$ via $W(\mathbf{x}_0,\mathbf{x}_i,\mathbf{z}_0,\mathbf{z}_i,\mathbf{h}) = \prod_{c=1}^{C} W_c(x_{0c},x_{ic},h_c)\cdot\prod_{d=C+1}^{C+D} W_d(z_{0d},z_{id},h_d)$. In the latter equation, $W_c(.)$ is the weighting function for continuous regressors, $W_d(.)$ is the weighting function for discrete regressors, x_{0c},x_{ic} denote values of continuous regressors, z_{0d},z_{id} denote values of discrete regressors, and h_c,h_d are smoothing parameters for continuous/discrete variables. As weighting function for continuous regressors, we use a second-order Gaussian kernel, the discrete regressors are weighted by the corresponding kernels of Li and Racine, described in [3].

As this nonparametric approach is based on locally approximating the true function with a linear function, it is called local linear estimation. The discrete regressors enter the minimization calculus only through the generalized product kernel but not through the quadratic term. Thus, the discrete covariates are included in a local constant fashion and, in contrast to the estimated partial effects $\widehat{\mathbf{b}}_1(\mathbf{x}_0,\mathbf{z}_0)$ for continuous variables, their partial effects are not estimated.

It is a nice feature of the local constant estimator that it delivers information on the relevance of regressors. If the bandwidth approaches infinity (its upper bound), the corresponding continuous (discrete) covariate has negligible or even no impact on the estimation. In case of local linear estimation, a bandwidth that is huge in relation to the range of the corresponding regressor indicates ceteris paribus a linear relationship between the regressors considered and the explanatory variable. Hence, local linear estimation provides information about the functional form without the need to plot the conditional mean function. This information can also be used to explore suitable parametric specifications. For discrete covariates an estimated bandwidth close to zero (the lower bound) indicates that the nonparametric regression is done almost separately for each level of the corresponding discrete covariate, in analogy to the classical frequency-approach, e.g., [8, Chap. 3].

For comparing the in-sample fit of the parametric and nonparametric specifications we consider the Pseudo-R^2

$$PR^2 = \left(\widehat{Corr}(\mathbf{y},\widehat{\mathbf{y}})\right)^2. \tag{9.3}$$

Hence, for each specification we analyze the (linear) relationship between observed and fitted values that are contained in the n-dimensional vectors $\mathbf{y}$ and $\widehat{\mathbf{y}}$.

As our first step of validating the parametric benchmark model we conduct the nonparametric misspecification test of Hsiao et al. (see [3]). Its pair of hypotheses is given by

$$H_0 : P(E[y_i|\mathbf{x}_i,\mathbf{z}_i] = f(\mathbf{x}_i,\mathbf{z}_i,\beta)) = 1 \quad \text{for some } \beta,$$
$$H_1 : P(E[y_i|\mathbf{x}_i,\mathbf{z}_i] = f(\mathbf{x}_i,\mathbf{z}_i,\beta)) < 1 \quad \text{for all } \beta.$$

Let the u_i's denote the errors of the parametric benchmark model. The test statistic T is based on estimating $E[u_i E[u_i|\mathbf{x}_i,\mathbf{z}_i]g(.)]$ and is consistent and asymptotically normal under quite general conditions. The test checks for remaining systematic information in the residuals of the parametric specification by running a nonparametric regression of these residuals on the covariates. Since the asymptotic distribution is known to kick in in rather large samples, `np` provides different bootstrapped versions of the test statistic.

As our second step of validation, the parametric benchmark specification is compared to a fully nonparametric model with respect to prediction performance. Bandwidths are chosen either by cross-validation or by the corrected Akaike criterion of Hurvich et al. (see [4]). We construct hold-out data for this comparison of the prediction performance by randomly splitting the sample in two parts. The first n_1 observations are used for estimating all specifications while the remaining $n_2 = n - n_1$ hold-out observations serve as validation-subsample used to compare the predicted values to the observed values. In the following, we replicate this step in a repeated random sampling Monte Carlo approach $B = 10{,}000$ times and compute for each replication the average squared error of prediction ($ASEP$) for measuring prediction performance

$$ASEP = \frac{1}{n_2}\sum_{i=1}^{n_2}(\widehat{y}_i - y_i)^2 \tag{9.4}$$

using the n_2 hold-out observations.

As described in Sect. 9.1 we then compute the percentage of the number of times when the parametric benchmark specification exhibits a lower $ASEP$ than the alternative nonparametric specification (and thus has the better prediction performance). Second, we compare the empirical distribution function of the $ASEP$s to see whether one specification dominates the other. This can be done by testing whether the difference in the mean of the $ASEP$s is significant or by comparing the distributions of the $ASEP$s for both model specifications.

9.3 Model Visualization and Validation Using `relax`

`relax` is an acronym for “R Editor for Literate Analysis and lateX” and is intended for combining report writing and statistical analysis. In the following, we especially use the `slider` function that allows interactive graphs by using the definition of sliders that can be dragged by left-clicking and holding the mouse button down. Changing the slider position immediately changes the results on the screen. For example, consider conditional scatter plots of two variables, where the slider position determines the value of the third variable. This is especially helpful in a regression setting of discrete and continuous covariates as one can plot the response variable against continuous covariates for the observations for different categories of discrete covariates determined by a slider. The `relax` package can be used for the exploration of the data structures, to detect relationships in the data that are not obviously visible. For example, these graphs can help to explore the inclusion of interactions in a parametric framework.

The `slider` function also allows the definition of buttons that are not moved, but just pressed, as is shown in the following example for a quick analysis of outliers.

```
(01) dataset <-
(02) axis.horizontal.column <- 1
(03) axis.vertical.column <- 2
(04) axis.horizontal <- dataset[, axis.horizontal.column]
(05) axis.vertical <- dataset[, axis.vertical.column]
(06) relax.plot <- function(...){
(07) plot(axis.horizontal, axis.vertical, pch = 20)
(08) if(slider(obj.name = "pick.obs") == 1){
(09)    screen.click <- locator(1)
(10)    temp <- (screen.click$x - axis.horizontal)^2 +
(11)    (screen.click$y - axis.vertical)^2
(12)    observation <- dataset[temp == min(temp),]
(13)    print(observation)
(14)    slider(obj.name = "pick.obs", obj.value = 0)
(15)    }}
(16) slider(relax.plot, but.functions = function(...){
(17) slider(obj.name = "pick.obs", obj.value = 1); relax.plot()}
(18) , but.names = c("select observation"))
(19) slider(obj.name = "pick.obs", obj.value = 0)
(20) relax.plot()
```

Code lines (01) to (03) are the only lines that need input from the user. The only prerequisite for the analysis is that the variables are contained in the same data set, saved as a matrix or a data frame where each row (column) corresponds to one observation (variable). In line (01), one has to include the name of the data set right after the arrow, while in line (02) one has to specify the column number of the variable that is intended to be on the horizontal axis of the scatter plot. Line (03) is equivalent to line (02) for the vertical axis. Lines (04) and (05) are convenient to simplify some of the following queries.

Lines (06) to (15) contain the information for plotting and printing the configuration of the selected observation, while lines (16) to (19) contain the configuration of the `slider` function. The last line (20) is simply a call to the plotting function defined in line (06) to start the graphical analysis. Line (07) contains the command for the scatter plot (we omit the additional code snippets necessary for improved graphs). Line (08) starts the if-query, whether the button "pick.obs" is pressed. Line (17) defines that if this button is clicked, it gets a value of 1 assigned, thus at the end of the if-query at line (14), the button is set to 0 again. Line (09) uses the `locator` function of the automatically loaded `graphics` package, where the coordinates of the point that is clicked in the scatter plot are saved as "screen.click$x" and "screen.click$y". These coordinates are used to determine the Euclidean distance in lines (10) and (11) between the clicked point and each observation of the data set. Thus, the configuration (i.e., the row of the data set) of the observation with the minimal Euclidean distance — see line (12) — is printed out in line (13). Line (15) closes the if-query for whether the button is pressed as well as the "relax.plot" function that generates the graphical setup.

Lines (16) to (18) contain the definition of the button "select observation" that is denoted as "pick.obs" for calling it within a function as in line (08). Hence, after pushing the button "select observation", the object "pick.obs" is set to 1. Line (19) is simply included for clean programming, as the button shall have the value 0 assigned before it is pressed. A second example that demonstrates the use of sliders is contained in the R appendix.

9.4 Beauty and the Labor Market Revisited

Referring to the seminal work of Hamermesh and Biddle (see [1], hereafter HB) on the impact of looks on earning, we use the data available on Hamermesh's homepage (`http://www.eco.utexas.edu/faculty/Hamermesh/`) to examine the validity of a parametric specification and corresponding inferences. We have $n = 1{,}260$ observations on the following variables (see HB): `wage`: hourly wage in US-$, `educ`: education in years, `fem`: dummy variable=1 if the observation is from a female, `look`: the sample values of this ordered categorical variable range from "1" (strikingly handsome), "2" (above average), "3" (average), "4" (below average), to "5" (homely). As suggested by HB, we assign the extreme categories to the nearest category, respectively. From this we also calculate the dummy variables `below` (`above`), which is equal to 1 if the respective looks are below (above) average.

In Table 9.1 and Figure 9.1 we provide numerical and graphical descriptions of the variables `wage`, `log(wage)`, `educ`, `look`, and `fem` and their respective relationships. Following HB we will use `log(wage)` as response variable in the regression analysis.

From our graphical and descriptive diagnostics we are careful to note that there is one quite extreme observation of a female with 13 years of education and below-average looks who has the maximum hourly wage of approximately $78. This observation, which is also highlighted in all of the residual diagnostics contained in the `plot()` command for all of the following linear regressions, is a "good" leverage point as trimming it does not alter the regression results reported below. To investigate potential leverage points we can also use the code of the relax example in Sect. 9.3. If we link the data

Table 9.1 Descriptive statistics generated with function `basicStats()` from package `fBasics`

	`wage`	`log(wage)`	`educ`
Minimum	1.02	0.02	5
Maximum	77.72	4.35	17
1. Quartile	3.71	1.31	12
3. Quartile	7.69	2.04	13
Mean	6.31	1.66	12.56
Median	5.30	1.67	12
Variance	21.7216	0.3534	6.8879
Stdev	4.6606	0.5945	2.6245
Skewness	4.8077	0.0831	-0.3713
Kurtosis	50.9277	0.4196	0.8735

set of HB in line (01) and enter the column number of `educ` in line (02) and that of `wage` in line (03), we get a scatter plot of `wage` on `educ` and clearly see that potential outlier. Clicking "select observation" and afterwards on the observation with an hourly wage of more than 70 US-$, produces the values of all variables for the corresponding row in the data set printed in the R console, without sorting and scrolling through the data set of 1,260 observations. This tool is especially useful to check whether extreme observations for one or more variables have certain similarities concerning the other variables in the data set.

In Table 9.2 we report the ordinary least squares (OLS) regression results and some diagnostics based on the wage equation (and nested smaller models)

$$\begin{aligned}\texttt{log(wage)} = &\beta_1 + \beta_2\texttt{above} + \beta_3\texttt{below} + \beta_4\texttt{fem} + \beta_5\texttt{educ} + \beta_6\texttt{educ}^2 + \\ &\beta_7\texttt{above}\cdot\texttt{fem} + \beta_8\texttt{below}\cdot\texttt{fem} + \beta_9\texttt{educ}\cdot\texttt{fem} + \beta_{10}\texttt{educ}^2\cdot\texttt{fem} + u. \quad (9.5)\end{aligned}$$

As the null of homoskedasticity cannot be rejected at any reasonable significance level, we report OLS standard errors. The regression results suggest that the most parsimonious (with respect to SC) specification $\mathbb{M}_{II}$ works reasonably and explains roughly 28% of the variation in `log(wage)`. We apply the test of [3], using their original configuration. The null model $\mathbb{M}_{II}$ produces the following p-values: 0.718 (asymptotic), 0.125 (iid bootstrap), and 0.138 (wild bootstrap). It is worth noting that by redoing the test with specification $\mathbb{M}_{IV}$ (favored by AIC), one obtains larger p-values throughout: 0.922 (asymptotic), 0.602 (iid bootstrap), and 0.654 (wild bootstrap).

A local linear estimator for mixed data is used to estimate the alternative nonparametric model where we apply the Li–Racine kernels for mixed data and a second-order Gaussian kernel for the continuous variables. The bandwidth choices reported in Table 9.3 are based on the corrected AIC of Hurvich et al. (see [4]). Note that the Pseudo-R^2 $PR^2 = 0.2805$ is only slightly larger than that of the parametric benchmark model. The bandwidth of `fem`

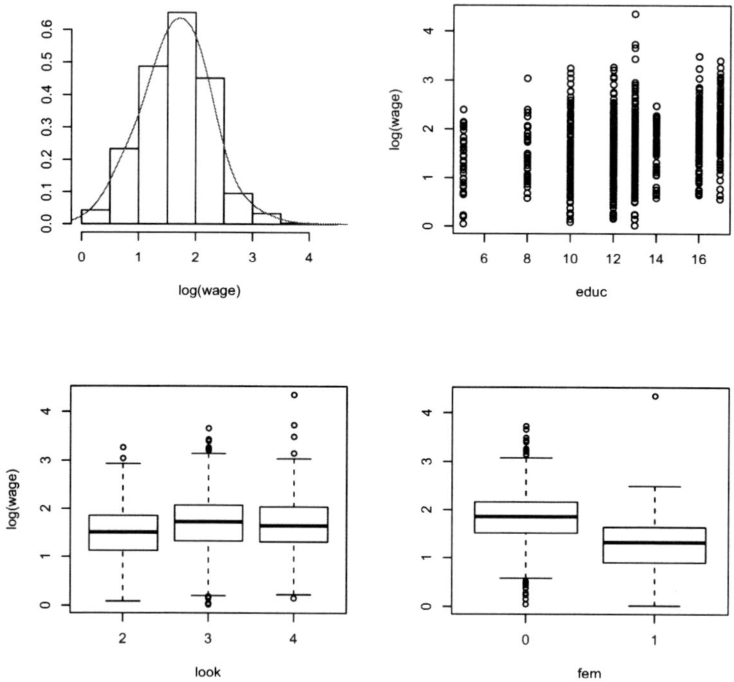

Fig. 9.1 Histogram and density estimate of `log(wage)`, plot of `log(wage)` against `educ`, and boxplots of `log(wage)` against `look` and `fem`.

is quite close to zero, hence the relationship between `wage` and `look` and `educ` is estimated for males and females separately.

In comparison to the nonparametric specification, the parametric model $\mathbb{M}_{II}$ ($\mathbb{M}_{IV}$) has a smaller *ASEP* for 6,156 (6,993) out of 10,000 replications given a splitting proportion of 90:10. The mean of the *ASEP*s for the parametric benchmark specification is significantly lower than the corresponding mean for the nonparametric specification, as a p-value below 0.0001 for the paired t-test reveals. In sum, the parametric benchmark specification $\mathbb{M}_{II}$ is successfully validated by both, the misspecification test and the prediction simulation.

To further investigate the prediction simulation results the use of `relax` is based on the example in Sect. 9.3. Here, we first generate a scatterplot for the 10,000 *ASEP*s of one specification against the other. An additional bisecting line helps to see whether for one or more of the 10,000 replications one specification predicts remarkably good/bad. Thus, this replication can be selected by clicking on it and can be further analyzed. For example, we can check whether for this replication only a few observations in the corresponding prediction sample are responsible for the high *ASEP*s or whether the prediction performance is poor for all observations for one of the specifications. For the

Table 9.2 OLS regression outputs for equation (9.5). OLS standard errors in parentheses

	$\mathbb{M}_I$	$\mathbb{M}_{II}$	$\mathbb{M}_{III}$	$\mathbb{M}_{IV}$	$\mathbb{M}_V$
`(const)`	1.8734 (0.0223)	1.6819 (0.1886)	1.7053 (0.1891)	1.6394 (0.2271)	1.6500 (0.2276)
`above`	−0.1774 (0.0471)	−0.1562 (0.0450)	−0.1671 (0.0567)	−0.1584 (0.0450)	−0.1703 (0.0567)
`below`	−0.0165 (0.0337)	−0.0577 (0.0323)	−0.0962 (0.0402)	−0.0544 (0.0323)	−0.0887 (0.0403)
`fem`	−0.5427 (0.0316)	−0.5420 (0.0301)	−0.5803 (0.0403)	−0.4451 (0.4026)	−0.4453 (0.4059)
`educ`		−0.0371 (0.0308)	−0.0393 (0.0309)	−0.0196 (0.0374)	−0.0204 (0.0374)
`educ`^{2}		0.0041 (0.0013)	0.0042 (0.0013)	0.0030 (0.0015)	0.0030 (0.0015)
`above` · `fem`			0.0344 (0.0930)		0.0364 (0.0932)
`below` · `fem`			0.1076 (0.0670)		0.0956 (0.0673)
`educ` · `fem`				−0.0472 (0.0657)	−0.0512 (0.0661)
`educ`^{2} · `fem`				0.0030 (0.0027)	0.0031 (0.0027)
PR^2	0.2006	0.2772	0.2787	0.2808	0.2820
$\bar{R}^2$	0.1987	0.2744	0.2747	0.2768	0.2768
AIC	1992.12	1869.18	1870.58	1866.91	1868.88
SC	2017.82	1905.15	1916.83	1913.16	1925.41

Table 9.3 Estimated and maximal bandwidth for mixed nonparametric regression using `np`.

Variable	Estimated bandwidth	Maximum bandwidth
`fem`	0.004250	1
`look`	0.152078	1
`educ`	4.917089	∞

data set of HB, we find no extreme replications such that the *ASEP* of one specification is several times as large as that of the other specification. The analysis of the prediction performance shows that both specifications deliver predictions of a quite similar quality with a slight advantage for the parametric specification. Hence, the parametric specification seems to capture the nonlinearity for the conditional expectation of the wages quite well.

Going beyond analyzing the conditional mean, further investigations on model validity may be directed to (i) whether OLS is the most accurate estimator for the central tendency of the conditional distribution and (ii) whether this relationship is stable across the conditional distribution. Both issues can be addressed by using quantile regression methods (e.g., [5]). This method is implemented in package `quantreg` by Koenker (see [6]). Some preliminary ϑ-quantile regression ($0 < \vartheta < 1$) results generated with function `rq()` are displayed in Table 9.4 and Fig. 9.2 (using model $\mathbb{M}_{//}$ selected by SC).

From these results we observe that the curvature of the impact of `educ` changes across the conditional distribution of `log(wage)`. Clearly, in the lower part the relationship is almost linear, whereas in the middle and upper part we find the previously observed quadratic relationship.

Table 9.4 Coefficient estimates for ϑ-regression quantiles applied to beauty data. Standard errors in parentheses generated from iid bootstrap

Covariates	0.1	0.2	0.3	0.4	0.5	0.6	0.7	0.8	0.9
`(const)`	0.937	1.234	1.161	1.481	1.785	1.956	2.324	2.381	2.457
	(0.382)	(0.262)	(0.283)	(0.250)	(0.268)	(0.222)	(0.169)	(0.229)	(0.295)
`above`	−0.079	−0.110	−0.165	−0.180	−0.185	−0.184	−0.211	−0.180	−0.159
	(0.091)	(0.063)	(0.068)	(0.060)	(0.064)	(0.053)	(0.040)	(0.055)	(0.070)
`below`	−0.061	−0.042	−0.050	−0.111	−0.082	−0.019	−0.053	−0.036	−0.081
	(0.065)	(0.045)	(0.048)	(0.043)	(0.046)	(0.038)	(0.029)	(0.039)	(0.050)
`fem`	−0.533	−0.574	−0.531	−0.506	−0.510	−0.563	−0.550	−0.572	−0.536
	(0.061)	(0.042)	(0.045)	(0.040)	(0.043)	(0.035)	(0.027)	(0.037)	(0.047)
`educ`	0.000	−0.019	0.011	−0.022	−0.057	−0.064	−0.105	−0.103	−0.073
	(0.062)	(0.043)	(0.046)	(0.041)	(0.044)	(0.036)	(0.028)	(0.038)	(0.048)
`educ`^{2}	0.002	0.003	0.002	0.003	0.005	0.005	0.007	0.007	0.006
	(0.003)	(0.002)	(0.002)	(0.002)	(0.002)	(0.001)	(0.001)	(0.002)	(0.002)

In addition, Figure 9.2 reveals that based on specification $\mathbb{M}_{//}$ the estimated median wage of average-looking males almost coincides with the estimated $\vartheta = .9$-quantile wage of average-looking females. It is worth noting that the parameter estimates for `fem` are quite stable across quantiles. The pure location shift null hypothesis cannot be rejected at any reasonable significance level.

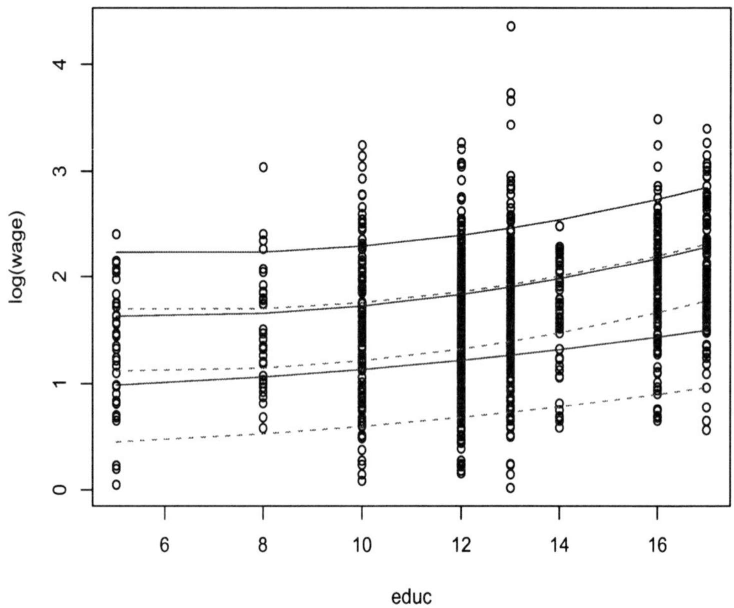

Fig. 9.2 Conditional plot of `log(wage)` against `educ` with fitted values for ϑ-regression quantiles ($\vartheta = .1,.5,.9$) for average-looking females (dashed red) and males (solid blue).

References

1. Hamermesh D S, Biddle J E (1994) Beauty and the Labour Market. Amer Econ Rev 84: 1174–1194
2. Hayfield T, Racine J S (2009) np: Nonparametric kernel smoothing methods for mixed datatypes. R package vers. 0.30-3
3. Hsiao C, Li Q, Racine J S (2007) A consistent model specification test with mixed discrete and continuous data. J Econometrics 140: 802–826
4. Hurvich C M, Simonoff J S, Tsai C L (1998) Smoothing Parameter Selection in Nonparametric Regression using an Improved Akaike Information Criterion. J Roy Stat Soc B 60: 271–293
5. Koenker R (2005) Quantile Regression. Cambridge University Press, Cambridge
6. Koenker R (2009) quantreg: Quantile Regression. R package vers. 4.38
7. Li Q, Racine J S (2004) Cross-validated Local Linear Nonparametric Regression. Statist Sinica 14: 485–512
8. Li Q, Racine J S (2007) Nonparametric Econometrics: Theory and Practice. Princeton University Press, Princeton
9. Racine J, Li Q (2004) Nonparametric Estimation of Regression Functions with both Categorical and Continuous Data. J Econometrics 119: 99–130
10. R Development Core Team (2009) R: A Language and Environment for Statistical Computing. R Foundation for Statistical Computing, Vienna
11. Wolf H P (2009) relax: R Editor for Literate Analysis and lateX. R package vers. 1.2.1

Chapter 10
Fitting Multinomial Models in R: A Program Based on Bock's Multinomial Response Relation Model

David Rindskopf

Abstract Bock's model for multinomial responses considered contingency tables as consisting of two kinds of variables, sampling variables (that defined groups) and response variables. Contrasts among response variables were specified, and these were modeled as functions of contrasts among categories defined by the sampling variables. This neat separation into independent and dependent variables was not captured by general log-linear model programs, but fits well within the framework that most social scientists are familiar with. The model is framed to parallel the usual multivariate analysis of variance (MANOVA) model, so those familiar with MANOVA will find the multinomial model very natural. This chapter describes an R function to fit this model, and provides several examples.

10.1 Model

Bock [2, 3] developed a model for the analysis of categorical data that has two advantages over other common formulations. First, it directly mimics the structure of multivariate analysis of variance, so that researchers who are used to that way of specifying models will find it natural to move from the analysis of continuous variables to the analysis of categorical variables. Second, the model assumes the data are in the form of a contingency table where rows represent groups (possibly based on combinations of variables) and columns represent responses (possibly based on more than one response variable). This is a natural way of conceptualizing the data and model structure for many social scientists, who commonly think in terms of independent and

David Rindskopf
Educational Psychology Program, CUNY Graduate Center, New York, NY 10016, USA e-mail: drindskopf@gc.cuny.edu

H.D. Vinod (ed.), *Advances in Social Science Research Using R*, Lecture Notes in Statistics 196, DOI 10.1007/978-1-4419-1764-5_10,

dependent variables. An example (analyzed later) uses the following data from Agresti [1]:

Sample		Response		
Gender	Race	Democrat	Republican	Independent
male	white	132	176	127
male	black	42	6	12
female	white	172	129	130
female	black	56	4	15

The combinations of gender and race (in a 2×2 design) make up the sample groups, and the levels of party affiliation represent the response categories. As with logistic regression (and its extension to outcomes with more than two categories), each row of the table is first transformed to proportions, and then to the (natural) logarithm of the odds of being in each response category; the resulting table consists of multinomial logits.

The statistical model for the multinomial logits is $Z = K\Gamma T$, where Z is a matrix of expected values of the multinomial logits, K is the model matrix for the sample, Γ contains the parameters, and T is the model matrix for the response variables. Each row of Z contains a vector of logits, which are related to probabilities through the natural extension of logits in the binomial case:

$$\pi_i = \exp(z_i) / \sum_j \exp(z_j)$$

The single subscript i is used to represent the column of Z; the column may, in practice, represent a combination of levels of dependent variables. The model for the multivariate logits is parallel to the usual model for multivariate analysis of variance, in which columns of K would represent contrasts among the between-subject factors, and rows of T would represent contrasts among within-subject factors.

For the data set on political party affiliation, the matrix F of observed frequencies would be

$$\begin{bmatrix} 132 & 176 & 127 \\ 42 & 6 & 12 \\ 172 & 129 & 130 \\ 56 & 4 & 15 \end{bmatrix}$$

The K matrix for a saturated model, using effects coding, could be

$$\begin{bmatrix} 1 & 1 & 1 & 1 \\ 1 & 1 & -1 & -1 \\ 1 & -1 & 1 & -1 \\ 1 & -1 & -1 & 1 \end{bmatrix}$$

And the T matrix for a saturated model, using coding to represent two possible contrasts among response categories, might be

$$\begin{bmatrix} 1 & 1 & -2 \\ 1 & -1 & 0 \end{bmatrix}$$

To interpret the parameter estimates, it is easiest to conceptualize the matrix Γ of parameters as the effect of a column of K on a row of T. This can be visualized in the following table (modeled after Bock [3, Table 8.1-10, p. 537]), which is based on the coding above:

	Party vs. indep	Dem vs. Repub
constant	γ_{11}	γ_{12}
gender	γ_{21}	γ_{22}
race	γ_{31}	γ_{32}
g × r	γ_{41}	γ_{42}

10.2 Program Code

The program code (on the website for this book) creates a function called mqual that implements in a straightforward manner the equations in Bock [2, 3]. I have omitted some features, such as the ability to include structural zeros. These do not occur frequently in common practice, so I chose instead to keep the program simpler. If one has structural zeros, it is easiest to reframe the problem so that all variables are response variables, set $K = [1]$, and omit cells with structural zeros from the data (see Rindskopf [5], for more information on this and other issues in fitting nonstandard models). Two versions of the function are given. The first version does not print labels for either the frequency table, the parameter matrix, or the contrast matrices; the second version does all of these if labels are put on the data and contrast matrices (this is illustrated in some of the examples below).

10.3 How to Use the mqual Function

To use the mqual function, one must first define three matrices: F, K, and T. (They need not have these names, but must be provided in that order to the mqual function.) F contains the observed frequencies in the form of a rectangular table. K contains the model for the samples (independent variables), and T contains the model for the responses. Note that K usually contains a column of 1s, while T never contains a row of 1s. F, K, and T must be matrices, not data frames. Once these are defined, the program is invoked by the command mqual(F, K, T).

Output is automatically displayed on the R console. It consists of the observed frequencies, observed and expected (under the model) row proportions, sample and response design matrices, Pearson and likelihood-ratio chi-square tests (with df and p values), and parameter estimates with their standard errors and standardized values (parameter estimates divided by standard errors). In this chapter, only selected output is presented for most examples.

10.4 Example 1: Test of Independence

The 2×2 table analyzed was the actual result of the first test of penicillin, which was done using eight mice that were infected with a deadly bacterium. The four treated with penicillin all survived, while all those not given penicillin died.

10.4.1 Input

```
ex.f <- matrix(c(4,0,0,4),nrow=2,byrow=T)
ex.k <- matrix(c(1,1),nrow=2)
ex.t <- matrix(c(1,0),nrow=1)
mqual(ex.f, ex.k, ex.t)
```

10.4.2 Output

```
...

Degrees of Freedom              =  1
Pearson Fit Statistic           =  8
Probability(Prsn)               =  0.004677735
Likelihood Ratio Fit Statistic  =  11.09035
Probability(LR)                 =  0.0008677788

...
```

10.5 Example 2: Effect of Aspirin on Myocardial Infarction (MI)

This data set comes from the first large randomized trial to determine whether aspirin reduced the likelihood of MI (heart attack). The outcomes were fatal MI, nonfatal MI, or no MI. The contrasts in the T matrix were constructed to determine whether aspirin (i) reduced the probability of MI, and/or (ii) reduced the fatality rate among those that had an MI. The latter looks promising from the descriptive statistics, but is not significant due to lack of power.

Treatment	Fatal MI	Nonfatal MI	No MI
Aspirin	5	99	10933
Placebo	18	171	10845

10.5.1 Input

```
ex.f <- matrix(c(5, 99, 10933, 18, 171, 10845),nrow=2,byrow=T)
ex.k <- matrix(c(1,1),nrow=2)
ex.t <- matrix(c(-1, -1, 2, 1, -1, 0),nrow=2, byrow=T)
mqual(ex.f, ex.k, ex.t)

ex.k2 <- matrix(c(1,1,1,-1),nrow=2,byrow=T)
mqual(ex.f, ex.k2, ex.t)
```

10.5.2 Output from Saturated Model

```
Observed Frequencies
  5  99 10933
 18 171 10845

Observed Row Proportions
 0.0004530217 0.008969829 0.9905771
 0.0016313214 0.015497553 0.9828711

Parameter Estimates
 1.9121173 -1.3092434
 0.1536367 -0.1835975
```

```
Standard Errors
 0.04348046 0.1302652
 0.04348046 0.1302652

Parameters/SE
 43.976477 -10.050597
  3.533466  -1.409413
```

10.6 Example 3: Race × Gender × Party Affiliation

This data set was discussed in the first section of this article. The obvious contrasts are constructed in the K matrix: Intercept, main effect of gender, main effect of race, and gender × race interaction. In T, the rows are constructed to see whether race or gender is related to (i) the probability of belonging to a major party rather than being independent, and (ii) the probability of being a Democrat (rather than a Republican), among those in a major party.

10.6.1 Input

```
# data from Agresti (2002, p. 303) (also p. 339 in 1st Ed)
# number of people in each party, by group

dem <- c(132, 42, 172, 56)
rep <- c(176, 6, 129, 4)
ind <- c(127, 12, 130, 15)

#  create labels to print table

gender <- c('m', 'm', 'f', 'f')
race   <- c('w', 'b', 'w', 'b')

#  put into data frame, print

party.id <- data.frame(gender, race, dem, rep, ind)
party.id

#  create matrices of frequencies, sample design, response design

party.f <- cbind(dem,rep,ind)

const <- c(1,1,1,1)      #  intercept
```

```
ge <- c(-1,-1,1,1)        # gender
ra <- c(1,-1,1,-1)        # race
ge.ra <- ge * ra          # gender x race

party.k <- as.matrix(cbind(const,ge,ra,ge.ra))

d.r <- c(1,-1,0)          # Dem vs Rep for party members
dr.i <- c(-1,-1,2)        # Party member vs independent

party.t <- rbind(d.r,dr.i)

rownames(party.f) <- c("male white  ",
                       "male black  ",
                       "female white",
                       "female black")

colnames(party.f) <- c("Dem", "Rep", "ind")

rownames(party.t) <- c("DR v I", "D v R")

colnames(party.k) <-
    c("intcpt","male","white","m.w")

mqual(party.f, party.k, party.t)
```

10.6.2 Output

```
Observed Row Proportions
                   Dem        Rep       ind
male white   0.3034483 0.40459770 0.2919540
male black   0.7000000 0.10000000 0.2000000
female white 0.3990719 0.29930394 0.3016241
female black 0.7466667 0.05333333 0.2000000

Parameters/SE
            DR v I      D v R
intcpt  1.11308417  6.5825656
male    0.61310442 -1.8211817
white   0.07654031 -6.5825656
m.w    -0.44007291  0.1690993
```

10.7 Nonstandard Loglinear Models

Consider the following data, showing a child's preference for presidential candidate conditional on his or her mother's and father's preference:

Mother's preference	Father's preference	Child's preference	
		Johnson	Goldwater
Johnson	Johnson	256	18
	Goldwater	14	7
Goldwater	Johnson	20	5
	Goldwater	45	93

One natural model is the main effects model that includes the effect of both father's and mother's preference on the child's preference for Johnson or Goldwater. The sample model matrix K for this model is

$$\begin{bmatrix} 1 & 1 & 1 \\ 1 & 1 & 0 \\ 1 & 0 & 1 \\ 1 & 0 & 0 \end{bmatrix}$$

The response model matrix T is

$$\begin{bmatrix} 1 & 0 \end{bmatrix}$$

Although this model fits well ($LR = .041$, $df = 1$, $p = .839$), there is another hypothesis of interest: Is the influence of mother's choice on the child's choice the same as that of the father? In order to test this, we fit a model with columns 2 and 3 of K added together; this produces a nonstandard loglinear model (Rindskopf [5]).

The sample model matrix is now

$$\begin{bmatrix} 1 & 2 \\ 1 & 1 \\ 1 & 1 \\ 1 & 0 \end{bmatrix}$$

The fit of this model is excellent: $LR = 1.095$, $df = 2$, $p = 0.578$. The data are consistent with the theory that the parents have an equal effect on the child's preference.

10.8 Technical Details of Estimation Procedure

(This section may be skipped without loss of continuity.)

The section is a highly condensed summary of the estimation procedure described in more detail in Bock [3, pp. 520–528]. It is based on the full-rank form of the model for the multinomial logits,

$$Z = K\Gamma T \tag{10.1}$$

For sample group j, let the observed response frequencies be in a vector

$$\mathbf{r}_j = \left[r_{j1}\, r_{j2} \ldots r_{jm}\right]'$$

and the corresponding observed response probabilities be in a vector

$$\mathbf{P}_j = \left[P_{j1}\, P_{j2} \ldots P_{jm}\right]'$$

Then define the matrix

$$\mathbf{W}_j = \begin{bmatrix} P_{j1}\,(1-P_{j1}) & -P_{j1}\,P_{j2} & \ldots & -P_{j1}\,P_{jm} \\ -P_{j2}\,P_{j1} & P_{j2}\,(1-P_{j2}) & \ldots & -P_{j2}\,P_{jm} \\ \ldots & & & \\ -P_{jm}\,P_{j1} & -P_{jm}\,P_{j2} & \ldots & P_{jm}\,(1-P_{jm}) \end{bmatrix} \tag{10.2}$$

The first derivatives are

$$g(\Gamma) = \sum_{j=1}^{n} T(r_j - N_j P_j) \otimes K_j \tag{10.3}$$

The matrix of second derivatives is

$$-H(\Gamma) = -\sum_{j=1}^{n} N_j T W_j T' \otimes K_j K_j' \tag{10.4}$$

From current values of parameters $\hat{\Gamma}_i$ on iteration i (often all zeros for initial values), calculate trial logits $\left[\hat{Z}_{jk}^{(i)}\right] = Z_i = K\hat{\Gamma}_i T$ and then trial (estimated) probabilities $\hat{\mathbf{P}}_i = \mathbf{D_i}^{-1}\left[e^{\hat{Z}_{jk}^{(i)}}\right]$ where $\mathbf{D_i}$ is a diagonal matrix, each element of which is a sum of the elements in the rows of $\left[e^{\hat{Z}_{jk}^{(i)}}\right]$.

Then calculate the following adjustments, which are added to the current estimate of the parameter matrix Γ:

$$\delta_{\mathbf{i}} = H^{-1}(\hat{\Gamma}_i) g(\hat{\Gamma}_i) \tag{10.5}$$

The iterations continue until the corrections are small; normally 10 or fewer iterations are sufficient, and because computations are very fast I have fixed the number of iterations rather than test for convergence.

The variance–covariance matrix of the parameter estimates is the inverse of the negative of the matrix of second derivatives of the likelihood function:

$$V(\hat{\Gamma}) = H^{-1}(\hat{\Gamma}) \tag{10.6}$$

The estimated standard errors of the parameter estimates are the square roots of the diagonal elements of $V(\hat{\Gamma})$.

10.9 Troubleshooting and Usage Suggestions

The mqual function has no error trapping, so if you do something wrong the program (but not R) will fail without much information to help you diagnose the problem. The following tips are provided to help prevent such errors, and help diagnose them when they occur.

- Even though one cannot test a saturated model, it is often useful to fit this model. Examining the parameters divided by their standard errors (the last section of output) gives a good picture of which effects are and are not needed in the model. (As with regression, collinearity can create problems of interpretation, so some caution is needed.)
- The matrix K should have as many rows as the matrix F, and no more columns than rows.
- It is often useful to test the null logit model; that is, a model in which the matrix K consists only of a column of 1s. This provides a baseline for comparison with other models. If K is only a column of 1s, be sure it is a matrix and not a vector. This problem is likely to occur if you start with a fuller K matrix, and only select the first column as a new K matrix. Depending on how this is done, R may "think" that with just one column, you mean for K to be a vector. You may have to have an explicit "matrix" statement to prevent this.
- The K matrix should not have redundant columns. For typical designs this problem is simple to avoid: Enter a constant, and as many columns for each effect as there are degrees of freedom for that effect. For a grouping variable with two categories, such as gender, there is one degree of freedom; for a variable with five categories, four degrees of freedom. For interactions, take products of all main effect columns for each variable included in the interaction, just as in using regression.
- The T matrix should have as many columns as F, and one fewer row than columns.
- The T matrix should not have a row with 1s. T specifies contrasts among response categories, and because it models logits of probabilities, there is one less row in T than there are categories in the responses (because the probabilities add to 1).
- Observed zero frequencies can sometimes cause numerical problems, depending on the model that is fit. The mqual program will seldom crash due to zero frequencies, because it stops after 10 iterations. However, a

telltale sign of numerical problems is very large standard errors for one or more parameters.

References

1. Agresti, A.: Categorical data analysis, 2nd edn. Wiley, New York (2002)
2. Bock, R.D.: Estimating multinomial response relations. In: R.C. Bose, et al. (eds.) Essays in probability and statistics. University of North Carolina Press, Chapel Hill (1970)
3. Bock, R.D.: Multivariate statistical methods in behavioral research. McGraw-Hill, New York (1975)
4. Bock, R.D., Yates, G.: MULTIQUAL: Log-linear analysis of nominal or ordinal qualitative data by the method of maximum likelihood. National Educational Resources, Inc., Chicago (1973)
5. Rindskopf, D.: Nonstandard log-linear models. Psychological Bulletin **108**, 150–162 (1990)

Chapter 11
A Bayesian Analysis of Leukemia Incidence Surrounding an Inactive Hazardous Waste Site

Ronald C. Neath

Abstract In this chapter we consider a subset of the data analyzed by Waller, Turnbull, Clark, and Nasca (*Case Studies in Biometry* 1994), concerning incidence of leukemia cases in an area surrounding the GE Auburn hazardous waste site in Cayuga County in upstate New York. The data consist of exposed population and leukemia cases by census block for the five-year period from 1978 to 1982, and the goal of our analysis is to quantify the extent to which close proximity to the hazardous waste site increases risk of contracting leukemia. We follow roughly the methodology of Wakefield and Morris (*JASA* 2001), who utilized a location-risk model embedded in a standard disease-mapping framework to analyze incidence of stomach cancer in relation to a municipal solid waste incinerator on the northeast coast of England. We describe in detail the three-stage Bayesian hierarchical model, and the selection of prior distributions for the model parameters. A major emphasis of this chapter will be on the use of R and WinBUGS, and the R2WinBUGS interface between them, in conducting the data analysis.

11.1 Introduction

In this chapter we undertake an analysis of a subset of the data originally considered by [11], concerning incidence of leukemia cases in an area surrounding the GE Auburn hazardous waste site in Cayuga County in upstate New York. The data consist of exposed population and leukemia cases by census block for the five-year period 1978–1982, and the goal of our analysis is to quantify the extent to which close proximity to the hazardous waste site increases risk of contracting leukemia. We follow roughly the methodology

Ronald C. Neath
Department of Statistics and CIS, Baruch College, City University of New York, New York, NY 10010, USA e-mail: ronald.neath@baruch.cuny.edu

H.D. Vinod (ed.), *Advances in Social Science Research Using R*, Lecture Notes in Statistics 196, DOI 10.1007/978-1-4419-1764-5_11,

of [10], who utilized a location-risk model embedded in a standard disease-mapping framework to analyze incidence of stomach cancer in relation to a municipal solid waste incinerator on the northeast coast of England.

Confidentiality requirements restrict the available data to regional summaries only; precise locations of cases and controls are unavailable. The location of each census block is taken to be its geographic centroid; we effectively allocate every case and every control in a census block to a single address at the regions geographic centroid.

We will refer to the hazardous waste site as the "putative point source" or just "point source," and the census blocks will be referred to as "areas" or "regions."

The remainder of this chapter is organized as follows. In Section 11.2, before any discussion of statistical modeling, we present numerical and graphical summaries of the data. In Section 11.3 we propose a model, essentially adapting the model of [10] to the present situation. Section 11.4 is concerned with the selection of prior distributions in our Bayesian hierarchical model, and Sect. 11.5 summarizes the data analysis itself, culminating in numerical and graphical summaries of the estimated posterior distributions. We make a few concluding remarks in Sect. 11.6.

11.2 Data Summaries

The data consist of location, population from 1980 census, and number of leukemia cases in the 5-year period 1978–1982 for the 30 census blocks whose geographic centroid lies within 10 km of the GE Auburn hazardous waste site. Distances from the site range from 0.33 km to 9.38 km, with an average distance of 2.58 km. Population counts range from 77 to 2422, with an average population of 1303. Disease counts range from 0.01 to 4.52. Disease counts take noninteger values because it was necessary to allocate cases for which the precise census block was unknown. The average number of leukemia cases per region is 1.11.

Figure 11.1 plots disease count per 1000 of population versus distance from putative point source. We see that there is at least a suggestion that disease rates might be higher close to the waste site. Three regions stand out from the others, one of which stands out from those, and all three are at distances well under the average.

11.3 The Model

Let $i = 1, \ldots, n$ index the census block for the $n = 30$ regions. For each i let

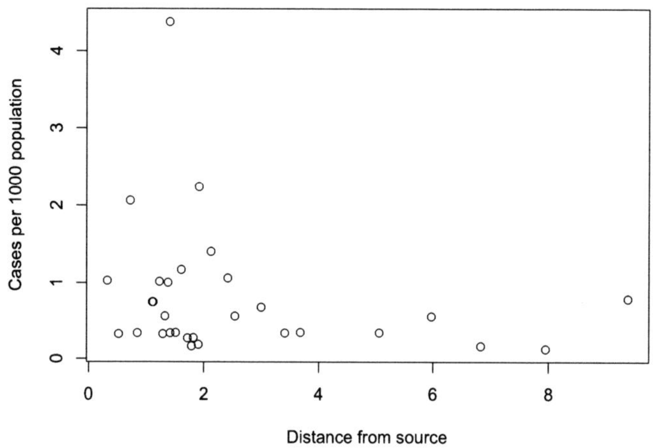

Fig. 11.1 Disease rate versus distance from point source.

$$\begin{aligned} Y_i &= \text{observed disease count in area } i \\ E_i &= \text{expected disease count in area } i \\ X_i &= \log(\text{population}) \text{ in area } i \\ d_i &= \text{distance from point source to centroid of area } i\,, \end{aligned}$$

and for each i and j let

$$d_{ij} = \text{distance between centroids of areas } i \text{ and } j\,.$$

We consider the following model, similar to that suggested for a similar problem by [10]. Assume, conditionally on the random effects U_i and V_i, whose distributions are specified below, that

$$Y_i \sim \text{ indep } \text{ Poisson}\left(E_i f(d_i;\alpha,\beta)\exp\{\eta_0+\eta_1 X_i+U_i+V_i\}\right)\,,$$

or equivalently, that the Y_i are independently Poisson distributed with means given by $E_i\lambda_i$ where

$$\log\lambda_i = \log f(d_i;\alpha,\beta)+\eta_0+\eta_1 X_i+U_i+V_i\,, \tag{11.1}$$

and

$$f(d;\alpha,\beta) = 1+\alpha\exp\left\{-(d/\beta)^2\right\} \tag{11.2}$$

is called the *location-risk function*. This completes the specification of the first stage of our three-stage hierarchical model. The second stage is specified by the distributions of the random effects $\mathbf{U} = (U_1,\ldots,U_n)^T$ and $\mathbf{V} = (V_1,\ldots,V_n)^T$.

We assume that $\mathbf{U}$ and $\mathbf{V}$ are independent multivariate normally distributed, specifically,

$$\mathbf{U} \sim \mathrm{N}_n(\mathbf{0}, \sigma_u^2 \mathbf{I}) \qquad \text{and} \qquad \mathbf{V} \sim \mathrm{N}_n(\mathbf{0}, \sigma_v^2 \mathbf{H}(\varphi))$$

where $\mathbf{H}(\varphi)_{ij} = \exp\left\{-\varphi d_{ij}\right\}$. Thus $U_i + V_i$ represents the region-specific effect associated with area i, which consists of independent (the U_i) and spatially correlated (the V_i) components. We complete the model specification with the third stage in Sect. 11.4, by attaching prior distributions to the unknown parameters $\alpha, \beta, \eta_0, \eta_1, \sigma_u^2, \sigma_v^2$, and φ.

We call this a *location-risk model embedded in a disease-mapping framework* because, but for the location-risk function f, this is the standard Poisson disease-mapping model with spatial and nonspatial extra-Poisson variability, as in Sect. 5.4 of [1]. Note that the location-risk model specified here might better be called a *distance*-risk model, since we assume that additional risk due to proximity to point source depends on distance only and not direction. More general *anisotropic* models, which do not make this assumption, are available but not appropriate for the present problem given the limited data at our disposal. The specific parametric form we chose is precisely that of [10], and is similar to that suggested by [2] in the context of a spatial point process model.

We can attach physical interpretations to the distance-risk function parameters α and β. Plugging $d = 0$ into (11.2) we find that the disease risk at the point source is equal to $1+\alpha$. Thus α can be thought of as the additional risk of leukemia faced by someone living with the waste site in his or her backyard. The additional risk at a distance d, as a proportion of that at the point source, is $e^{-(d/\beta)^2}$. Thus the distance at which the additional risk is reduced to 5% of its level at the point source itself is equal to $\beta\,(-\log(.05))^{1/2}$, or roughly $\beta\sqrt{3}$.

The covariate $X_i = \log(\text{population})$ of region i is included in the model based on the argument that, if census blocks are roughly the same geographic size, a high block population would indicate crowding, which seems a reasonable surrogate for deprivation in an area like upstate New York. Of course, some measure of the socioeconomic status of a region would be a preferable covariate (in the study of [10] they used the Carstairs index), but no such data were available for the present study. It is noted in [11] that the census blocks were originally constructed to have roughly equal populations. The 30 blocks in the study clearly do not (block populations range from 77 to 2422, with $\bar{x} = 1303$ and $s = 674$). Thus block population must indicate something about what has happened to an area in the years since the census block boundaries were established, though perhaps it is not entirely clear what.

The expected disease counts E_i are determined based on *internal standardization*, according to

$$E_i = \frac{(\text{Population in region } i) \times (\text{Total disease cases})}{\text{Total population}} .$$

A preferable approach would be to adjust for age and sex composition of the regions, but these data are not available.

The random effects U_i and V_i represent, respectively, the nonspatial and spatial extra-Poisson variability. We caution that, as noted by [10], the inclusion of random effects may dilute the effect of interest, namely, the distance-risk model f. We will revisit this issue in Sect. 11.4 in our discussion of the prior distributions for σ_u^2 and σ_v^2.

11.4 Prior Distributions

In this section we discuss the selection of prior distributions for the parameters $\alpha, \beta, \eta_0, \eta_1$ and hyperparameters σ_u^2, σ_v^2, and φ. In [10], the priors were informed by a preliminary mapping study. In the present study we have no such access to preliminary data, and thus our prior distributions must be based entirely on subjective judgment. We will assign informative priors where appropriate, and vague priors otherwise.

For the Poisson regression coefficients we adopt vague priors: take η_0 and η_1 to be independent and identically distributed as $\mathrm{N}(0, 1000)$.

Note that α is bounded from below by -1, and that zero is a "critical value" in the sense that $\alpha = 0$ corresponds to there being zero excess risk associated with the point source. A translated lognormal distribution seems a reasonable candidate for the prior on α. Recall that a lognormal random variable, say T, with $\mu = 0$ has equal probability of being less than 1 as of being greater than 1, and further has the property that $\Pr(r < T < s) = \Pr(1/s < T < 1/r)$ for any $1 < r < s$ (this follows from the distribution's symmetry about 0 on the log scale). Let us choose a prior distribution for α that enjoys this symmetry property. A sensible prior for α is defined by

$$\text{Let } a \sim \mathrm{N}(0,1), \quad \text{and let } \alpha = e^a - 1 .$$

Of course, $\sigma^2 = 1$ is no more or less arbitrary than any other hyperparameter value. This particular choice yields a prior probability of 0.95 that the excess risk at the point source falls between -86% and $+610\%$, which seems reasonable to us.

Recall the physical interpretation we attached to β in Sect. 11.3: The excess disease risk associated with the point source dips to 5% of its highest level at a distance of $\beta\sqrt{3}$ km. Let us suppose we are 95% certain that this value is smaller than 10, and also 95% certain that this value is larger than 1. Then the 5th and 95th percentiles of our prior distribution on β are equal to $1/\sqrt{3}$ and $10/\sqrt{3}$, respectively. A member of the gamma family of distributions that approximately matches these quantiles has a shape parameter of 2.5 and rate parameter of 1, that is, prior density satisfying $\pi(\beta) \propto \beta^{3/2}e^{-\beta}$. We will adopt this prior for our analysis.

For the variance components, [10] caution against the use of a vague prior that does not assign enough prior probability to very small values of σ_u^2 and σ_v^2, noting that the inclusion of random effects might dampen the effect in which our primary interest lies, namely, the distance-risk function parameters α and β. We will adopt these authors' recommendation and take as priors on the precisions σ_u^{-2} and σ_v^{-2} independent gamma distributions with shape parameter 0.5 and rate parameter .0005.

We will assign the spatial correlation parameter φ a prior distribution from the gamma family. Since it seems inappropriate for this prior to have a mode at 0, we wish to assign a shape parameter greater than 1. After some trial and error we settled on a gamma prior with shape parameter 3 and rate parameter 1. The 1st percentile of this distribution is 0.44, and $\varphi = 0.44$ corresponds to a spatial correlation of .01 at 10 km. For $\varphi > 0.44$, the spatial correlation is weaker.

11.5 Analysis

Let $\theta = (\alpha, \beta, \eta_0, \eta_1, \sigma_u, \sigma_v, \varphi)$ denote the vector of model parameters. A Bayesian data analysis is based on the posterior distribution with density $\pi(\theta|\mathbf{y})$ which, according to Bayes' rule, satisfies $\pi(\theta|\mathbf{y}) \propto g(\mathbf{y}|\theta)\pi(\theta)$, where $\pi(\cdot)$ denotes the prior density on θ and $g(\cdot|\theta)$ denotes the density of the observable data. In the present problem, the probability model for the observable data $\mathbf{Y} = (Y_1, \ldots, Y_n)$ is specified conditionally on the unobservable random effects $\mathbf{U}$ and $\mathbf{V}$. Let $g(\mathbf{y}|\mathbf{u}, \mathbf{v}, \theta_1)$ denote this conditional density, with $\theta_1 = (\alpha, \beta, \eta_0, \eta_1)$, and let $h(\mathbf{u}, \mathbf{v}|\theta_2)$ denote the joint density of the random effects $(\mathbf{U}, \mathbf{V})$, which depends on the parameters $\theta_2 = (\sigma_u, \sigma_v, \varphi)$. We can then write

$$g(\mathbf{y}|\theta) = \int\int g(\mathbf{y}|\mathbf{u}, \mathbf{v}, \theta_1)h(\mathbf{u}, \mathbf{v}|\theta_2)\, d\mathbf{u}\, d\mathbf{v}\,. \tag{11.3}$$

Recall that $g(\mathbf{y}|\mathbf{u}, \mathbf{v}, \theta_1)$ is a product of Poisson probabilities, and that $h(\mathbf{u}, \mathbf{v}|\theta_2)$ is the product of two multivariate normal densities. In any case, the integral in (11.3) cannot be solved analytically, and thus no closed form expression exists for the posterior density $\pi(\theta|\mathbf{y})$. In fact, even if $g(\mathbf{y}|\theta)$ could be evaluated, the posterior would still be analytically intractable in the sense that posterior moments and quantiles could not be solved for explicitly.

The usual remedy for this situation is to approximate the posterior by Monte Carlo simulation. In a *Markov chain Monte Carlo* (MCMC) Bayesian analysis, one simulates an ergodic Markov chain whose unique stationary distribution is given by the posterior density $\pi(\theta|\mathbf{y})$. The Metropolis–Hastings algorithm permits the simulation of such a chain for (in principle) any probability density whose form is known possibly up to an unknown normalizing constant. In a three-stage hierarchical model like that we are working with in the present problem, we might employ the following trick in our MCMC analysis. Define the joint "posterior" of the parameters and the random effects by

$$\pi(\boldsymbol{\theta}, \mathbf{u}, \mathbf{v}|\mathbf{y}) \propto g(\mathbf{y}|\mathbf{u}, \mathbf{v}, \boldsymbol{\theta}_1) h(\mathbf{u}, \mathbf{v}|\boldsymbol{\theta}_2) \pi(\boldsymbol{\theta}_1, \boldsymbol{\theta}_2) . \tag{11.4}$$

Explicit expressions are available for each term on the right-hand side. Thus the density for this target distribution is known up to a normalizing constant, and thus Metropolis–Hastings can be used to simulate an ergodic Markov chain with stationary density given by (11.4). Let $(\boldsymbol{\theta}^{(t)}, \mathbf{u}^{(t)}, \mathbf{v}^{(t)})$ for $t = 1, 2, \ldots$ denote the resulting chain. If the $\mathbf{u}^{(t)}$ and $\mathbf{v}^{(t)}$ terms are discarded, then $\boldsymbol{\theta}^{(t)}$ for $t = 1, 2, \ldots$ represents a Markov chain with stationary density $\pi(\boldsymbol{\theta}|\mathbf{y})$.

The construction of such a Markov chain from first principles requires a considerable amount of programming expertise. Fortunately an alternative exists, in the WinBUGS software of [7], that has made Bayesian data analysis methods available to researchers without such expertise. The WinBUGS user specifies a Bayesian hierarchical model, and the software simulates the desired Markov chain, returning MCMC approximations to the marginal posteriors of the individual model parameters.

Using WinBUGS we ran the Markov chain for 1000 updates, and discarded the realizations up to that point — this is sometimes called the *burn-in* method of selecting starting values. We then ran the chain for an additional 10^6 updates, but saved only every 10th draw to reduce the autocorrelation in our sample. Thus we have a total MCMC sample size of 10^5. The estimated posterior distributions of $\alpha, \beta, \eta_0, \eta_1, \sigma_u, \sigma_v, \varphi$ are summarized in Sect. 11.5.1.

11.5.1 Estimated Posteriors

Estimated posterior densities for $\eta_0, \eta_1, \alpha, \beta, \sigma_u, \sigma_v$ are given in Figure 11.2. Numerical summaries of the estimated posterior distributions are given in Table 11.1. The Monte Carlo standard error (MCSE) in the second column of Table 11.1 is an estimate of the expected error in the Monte Carlo approximation to the posterior mean. Note that the MCSE is *not* equal to the estimated standard deviation divided by the square root of the Monte Carlo sample size, as that naive rule would, in its failure to account for the autocorrelation in the MCMC sample, understate the true error. Instead we use the method of *nonoverlapping batch means* — the reader should refer to [5] or [6] for more details on MCMC standard error estimation.

The posterior mean of the η_1, the coefficient of the log-population in the Poisson regression component of our model, is positive, suggesting that area population may indeed be a reasonable surrogate for deprivation. However, a 90% credible region (the range from the 5th to the 95th percentiles of the

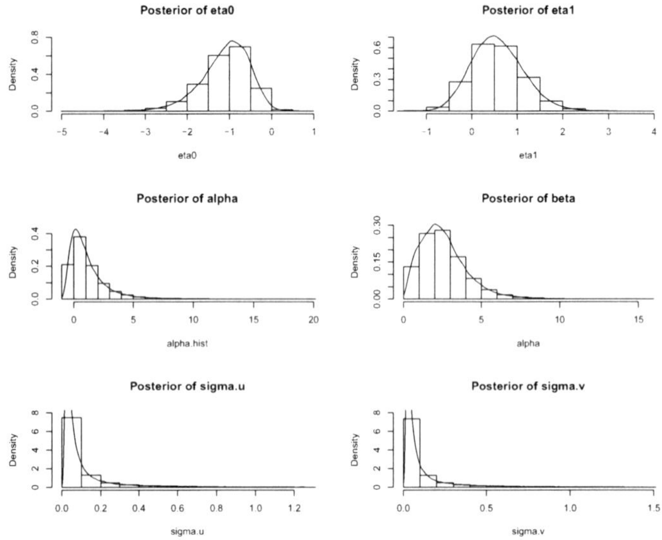

Fig. 11.2 Estimated posterior densities of model parameters.

estimated posterior) does cover zero, and thus it is by no means clear that population is an informative predictor of disease prevalence.

Table 11.1 Numerical summary of estimated marginal posteriors

	Mean	MCSE	Std Dev	5%ile	Median	95%ile
η_0	−1.10	.005	0.58	−2.16	−1.03	−0.28
η_1	0.57	.003	0.58	−0.32	0.54	1.57
α	1.21	.016	1.89	−0.46	0.72	4.48
β	2.58	.008	1.52	0.59	2.34	5.40
σ_u	0.10	.003	0.16	0.02	0.05	0.41
σ_v	0.11	.003	0.18	0.02	0.05	0.48
φ	3.00	.006	1.73	0.82	2.68	6.29

Note the severe right-skewness in the estimated posterior density of α. If we were to take the posterior mean of this distribution as an estimate $\hat{\alpha} = 1.21$, we would believe that the risk of leukemia more than doubles (221%) at close proximity to the hazardous waste site, relative to a location far away from the source (the posterior median leads to a considerably more conservative conclusion). Further, if we take the posterior mean of β as an estimate of that parameter, $\hat{\beta} = 2.58$, we would conclude that the excess risk of leukemia falls below one percent of the level at close proximity at a distance of 5.54 km from the point source.

Finally, we note that the data contain very little information about the variance parameters $\sigma_u, \sigma_v, \varphi$. The posterior distributions of σ_u and σ_v are very similar, and both are extremely close to their common prior. The estimated posterior of the spatial correlation parameter φ is essentially identical to its gamma(3,1) prior.

11.5.2 The Location-Risk Function

Figure 11.3 shows the posterior median value and individual 90% credible regions for the distance-risk function $f(d;\alpha,\beta)$, defined by (11.2), at $d = 0, 1, \ldots, 9$ km. A solid line connects the median values, and dashed lines connect the 5th and 95th percentile values. Consistent with our earlier observation, we see that the estimated risk at close proximity to the point source (corresponding to $d = 0$) is approximately double the baseline value, and reduces to no excess risk to speak of at a distance of about 5.5 km. Unfortunately, this graph further demonstrates the weakness of the inference that can be drawn from our limited data set. Just as the 90% credible interval for the excess risk parameter α included negative values, the 90% credible region for $f(d;\alpha,\beta)$ fails to preclude the possibility that disease risk is actually lessened by close proximity to a hazardous waste site.

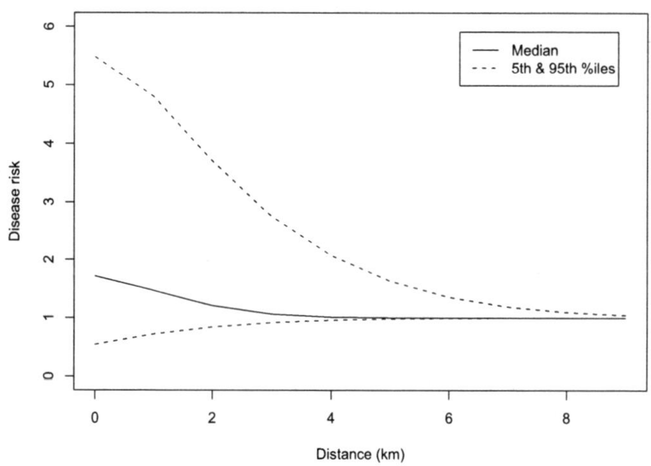

Fig. 11.3 Posterior of distance-risk function.

11.5.3 A Simplified Model

The inclusion of the random effects $U_i + V_i$ in the excess risk function (11.1), particularly the spatial component V_i, is controversial, and there are compelling theoretical and pragmatic arguments for removing those terms from the model. Note that in Poisson regression, unlike a Gaussian linear model, a random error term is not strictly necessary, as the Poisson parameter $E_i\lambda_i$ represents both the mean and variance in the number of cases in region i. Random effects can be included if the analyst wishes to accommodate the possibility of extra-Poisson variability in observed counts. Absent compelling evidence of extra-Poisson variability in the data, one could argue on the principle of parsimony that the region-specific random effects $U_i + V_i$ should not be included in our model. More pragmatically, there is concern that the inclusion of random effects might dilute the effect in which our primary interest lies, namely, that of the location-risk parameters α and β. We saw in Sect. 11.5.1 that the data contained very little information about the variance components σ_u and σ_v, which might be interpreted as an absence of compelling evidence that they are not identically zero. With the objective of sharpening our inference about α and β, we propose to refit the above model with the random effects $U_i + V_i$ excluded from the excess-risk function (11.1). Of course, this is precisely the model of Sects. 11.3 and 11.4, with $\sigma_u = \sigma_v = 0$.

Here we will make use of the R2WinBUGS package of [9], which provides an interface for running WinBUGS from R. Once the R2WinBUGS library is loaded, the user must specify the Bayesian hierarchical model (in WinBUGS syntax) in an external file, load the data, and define a rule for generating starting values for the MCMC. The R function `bugs()` then calls up WinBUGS, runs the simulations, and saves relevant summaries of MCMC output to the R session. In addition to numerical summaries of the estimated posterior distributions, which we will illustrate below, R2WinBUGS computes the MCMC convergence diagnostics of [4] and the Deviance Information Criterion (DIC) of [8]. A detailed summary of the latter objects is beyond the scope of this chapter. We note here that DIC is a popular Bayesian model selection criterion, and refer the interested reader to [8] for further information. For an interesting empirical study of the Gelman–Rubin Diagnostic and other MCMC convergence criteria, the reader is referred to [3].

The following display shows partial output of the `bugs()` function in the simplified (no random effects) model.

```
Inference for Bugs model at "C:/DOC~1/RNeath/...
 5 chains, each with 1e+05 iterations
  (first 50000 discarded),
 n.thin = 250
 n.sims = 1000 iterations saved

          mean   sd 2.5%  25%  50%  75% 97.5%
```

```
eta0      -1.1 0.5 -2.2 -1.4 -1.0 -0.7  -0.2
eta1       0.6 0.6 -0.4  0.2  0.6  0.9   1.7
alpha      1.2 1.6 -0.6  0.1  0.7  1.8   5.6
beta       2.6 1.5  0.4  1.5  2.5  3.4   6.0
```

The estimated posterior means, medians, and standard deviations for the regression parameters η_0 and η_1, as well as the location-risk parameters α and β, are essentially identical to those of the full (random effects) model, reported in Table 11.1. Thus our inference about the excess risk parameters appears not to depend on the inclusion or exclusion of extra-Poisson variability in the model for disease frequency.

11.6 Discussion

In this chapter we used a location-risk model embedded in a standard disease-mapping framework to analyze leukemia incidence data in a 10-km-radius area surrounding a hazardous waste site in upstate New York. From posterior inference in our three-stage hierarchical Bayesian model we found some indication, though far from overwhelming evidence, of increased disease risk at close proximity to the site. Monte Carlo approximations to the posterior distributions were computed by the WinBUGS software, and graphical and numerical summaries of the estimated posteriors were computed using R. Finally, we demonstrated the use of the R library R2WinBUGS, as an interface between those two software packages.

While it is somewhat disappointing that stronger conclusions could not be reached from the analysis, it is not surprising given the limited data. Leukemia is too rare a disease for statistically interesting patterns to emerge from five years of data on a population at risk of less than 40,000, in which only 33 cases were counted. Other limitations in the data include: (1) only areal summary data were available — with precise locations of cases and controls it is possible that more meaningful conclusions might have been reached; (2) the internal standardization used to calculate expected disease counts did not reflect the age and sex composition of the regions, possibly important information that was missing from available data; and (3) socioeconomic status might have been a useful covariate to include in the model, as a known confounder of disease prevalence, but was also unavailable.

Finally, we note that while we focused on Bayesian inference about the unknown parameters, from a public policy standpoint, there may be just as much interest in the prediction of unobserved random effects $U_i + V_i$, as this might indicate to which areas greater education and prevention efforts should be directed. Prediction of random effects in a Bayesian analysis is extremely straightforward. Recall from Sect. 11.5 that Monte Carlo simulation from the posterior $\pi(\theta|\mathbf{y})$ is actually accomplished by extracting the $\theta^{(t)}$ terms out of simulated draws from the "joint posterior" $\pi(\theta,\mathbf{U},\mathbf{V}|\mathbf{y})$, given by (11.4).

But it is also the case that the $(\mathbf{u}^{(t)}, \mathbf{v}^{(t)})$ terms define a Markov chain whose stationary distribution is the conditional distribution of the random effects given the data. Thus the ergodic average of the $u_i^{(t)} + v_i^{(t)}$ provides a Monte Carlo approximation to the mean of the predictive distribution of $U_i + V_i$.

Acknowledgements The author thanks Brad Carlin and Galin Jones for many informative discussions about Bayesian modeling and Markov chain Monte Carlo, respectively, and Lance Waller for providing the data.

References

1. Banerjee, S., Carlin, B.P., Gelfand, A.E.: Hierarchical Modeling and Analysis for Spatial Data. Chapman & Hall / CRC (2004)
2. Diggle, P.J.: A point process modelling approach to raised incidence of a rare phenomenon in the vicinity of a prespecified point. Journal of the Royal Statistical Society, Series A **153**, 349–362 (1990)
3. Flegal, J.M., Haran, M., Jones, G.L.: Markov chain Monte Carlo: Can we trust the third significant figure? Statistical Science **23**, 250–260 (2008)
4. Gelman, A., Rubin, D.B.: Inference from iterative simulation using multiple sequences. Statistical Science **7**, 457–472 (1992)
5. Jones, G.L., Haran, M., Caffo, B.S., Neath, R.: Fixed-width output analysis in Markov chain Monte Carlo. Journal of the American Statistical Association **101**, 1537–1547 (2006)
6. Jones, G.L., Hobert, J.P.: Honest exploration of intractable probability distributions via Markov chain Monte Carlo. Statistical Science **16**, 312–334 (2001)
7. Lunn, D.J., Thomas, A., Best, N., Spiegelhalter, D.: WinBUGS – a Bayesian modelling framework: Concepts, structure, and extensibility. Statistics and Computing **10**, 325–337 (2000)
8. Spiegelhalter, D.J., Best, N.G., Carlin, B.P., van der Linde, A.: Bayesian measures of model complexity and fit. Journal of the Royal Statistical Society, Series B **64**, 583–616 (2002)
9. Sturtz, S., Ligges, U., Gelman, A.: R2WinBUGS: A package for running WinBUGS from R. Juornal of Statistical Software **12**, 1–16 (2005)
10. Wakefield, J.C., Morris, S.E.: The Bayesian modeling of disease risk in relation to a point source. Journal of the American Statistical Association **96**, 77–91 (2001)
11. Waller, L.A., Turnbull, B.W., Clark, L.C., Nasca, P.: Spatial pattern analyses to detect rare disease clusters. In: N. Lange, L. Ryan, L. Billard, D. Brillinger, L. Conquest, J. Greenhouse (eds.) Case Studies in Biometry, pp. 3–23. John Wiley & Sons, Inc. (1994)

Chapter 12
Stochastic Volatility Model with Jumps in Returns and Volatility: An R-Package Implementation

Adjoa Numatsi and Erick W. Rengifo

Abstract In this chapter we estimate the stochastic volatility model with jumps in return and volatility introduced by [7]. In this model the conditional volatility of returns can not only increase rapidly but also persistently. Moreover, as shown by [8], this new model performs better than previous models presenting almost no misspecification in the volatility process. We implement the model coding the algorithm using R language. We estimate the model parameters and latent variables using FTSE 100 daily returns. The values of some of our estimated parameters are close to values found in previous studies. Also, as expected, our estimated state variable paths show high probabilities of jumps in the periods of financial crisis.

12.1 Introduction

Modeling equity returns remains an important topic in current financial literature. Even though the Black–Scholes model remains as one of the most used benchmarks in the industry, it has been shown that the results obtained from its main formula are biased. This is mostly due to two of its more important assumptions [14]: first, that stock prices follow a continuous path through time and that their distribution is lognormal, and second, that the variance of stock returns is constant [4]. In many empirical studies it has been found that asset returns' unconditional distributions feature a greater degree of kur-

Adjoa Numatsi
Department of Economics, Fordham University, Bronx, NY 10458, USA
e-mail: numatsi@fordham.edu

Erick W. Rengifo
Department of Economics, Fordham University, Bronx, NY 10458, USA
e-mail: rengifomina@fordham.edu

H.D. Vinod (ed.), *Advances in Social Science Research Using R*,
Lecture Notes in Statistics 196, DOI 10.1007/978-1-4419-1764-5_12,

tosis than implied by the normality assumption, and that volatility clustering is present in the data suggesting random changes in returns' volatility [5].

In an attempt to improve on the Black–Scholes model, both assumptions have been relaxed. There are models that can capture large movements in stock prices allowing for discontinuities in the form of jump diffusion models [12, 6], and models that allow for stochastic volatility [10, 16, 9]. Bates [2] and Scott [15] combined these two approaches and introduced the stochastic volatility models with jumps in returns. However, while it is clear that both stochastic volatility and jumps in returns are important components of the dynamics of stock prices, Eraker, Johannes and Polson [8] showed that the volatility process is misspecified. Similar results were found by [1], [3], and [13].

Empirical studies have shown that conditional volatility of returns increases rapidly, a feature that stochastic volatility with jumps in returns are not able to capture. Jumps in returns can generate large movement such as the crash of 1987, but the impact of a jump is temporary and dies out quickly. On the other hand, diffusive stochastic volatility models produce persistent volatility movements but since its dynamics are driven by a Brownian motion, volatility only increases gradually by small normally distributed increments. Given these findings there was a need to make a model that can create quick and persistent movements of the conditional volatility of returns. Duffie, Pan and Singleton [7] were the first to introduce models with jumps in both returns and volatility, and Eraker, Johannes and Polson [8] the ones who implemented it. As mentioned by [8], the estimation results showed that the new model with jumps in returns and jumps in stochastic volatility performed better than previous models presenting almost no misspecification in the volatility process.

The objective of this chapter is to implement the stochastic volatility model with jumps in returns and volatility using R. R is a free software heavily used in mathematics and statistics. For example, a recent econometrics text [17] emphasizes R and includes finance applications. R is a very robust programming language that has already many built-in packages that make it straightforward to use in many other disciplines like finance. However, to the best of our knowledge, there is no R package that deals with these type of models where jumps appear not only in returns but also in volatility. We expect to contribute with this program to the existing library of programs that can be used either by academics or practitioners alike.

The remaining chapter is organized as follows: Section 12.2 follows [8] to describe the stochastic volatility model with jumps in returns and volatility. Section 12.3 introduces the data set, the methodology followed to estimate the model parameters, and the structure of the R program. Finally in this section, we present the results of the model computed using the R program. Section 12.4 concludes and presents venues for future research.

12.2 The Stochastic Volatility Model with Jumps in Returns and Volatility

This section follows [8]. For additional information we refer the reader to this paper. The stochastic volatility model is a jump diffusion process with double jumps: a jump in returns and in volatility. These jumps arrive simultaneously with the jump sizes being correlated. According to the model, the logarithm of asset's price $Y_t = \log(S_t)$, solves

$$\begin{pmatrix} dY_t \\ dV_t \end{pmatrix} = \begin{pmatrix} \mu \\ \kappa(\theta - V_t) \end{pmatrix} dt + \sqrt{V_{t^-}} \begin{pmatrix} 1 & 0 \\ \rho\sigma_v & \sqrt{(1-\rho^2)}\sigma_v \end{pmatrix} dW_t + \begin{pmatrix} \xi^y dN_t^y \\ \xi^v dN_t^v \end{pmatrix} \tag{12.1}$$

where $V_{t^-} = \lim_{s\uparrow t} Vs$, W_t is a standard Brownian motion in R^2. The jump arrivals N_t^y and N_t^v are Poisson processes with constant intensities λ_y and λ_v. This model assumes that the jump arrivals are contemporaneous ($N_t^y = N_t^v = N_t$). The variables ξ^y and ξ^v are the jump sizes in returns and volatility, respectively. The jump size in volatility follows an exponential distribution $\xi^v \sim \exp(\mu_v)$ and the jump sizes in returns and volatility are correlated with $\xi^y|\xi^v \sim N(\mu_y + \rho_j\xi^v, \sigma_y^2)$.

Equation (12.1) presents the model in its continuous form. In order to estimate the model, Equation (12.1) is discretized obtaining

$$Y_{(t+1)\Delta} - Y_{t\Delta} = \mu\Delta + \sqrt{V_{t\Delta}\Delta}\varepsilon^y_{(t+1)\Delta} + \xi^y_{(t+1)\Delta}J^y_{(t+1)\Delta} \tag{12.2}$$

$$V_{(t+1)\Delta} - V_{t\Delta} = \kappa(\theta - V_{t\Delta})\Delta + \sigma_v\sqrt{V_{t\Delta}\Delta}\varepsilon^v_{(t+1)\Delta} + \xi^y_{(t+1)\Delta}J^v_{(t+1)\Delta} \tag{12.3}$$

where $J^k_{(t+1)\Delta} = 1(k = y, v)$ indicates a jump arrival. Jump times are Bernoulli random variables with constant intensities, $\lambda_y\Delta$ and $\lambda_v\Delta$. The distributions of the jump sizes remain the same. $\varepsilon^y_{(t+1)\Delta}$ and $\varepsilon^v_{(t+1)\Delta}$ are standard normal random variables with correlation ρ. The time-discretization interval Δ is assumed to be one day. As usual, it is possible that the time-discretization procedure introduces a discretization bias. However, for this particular case, [8] provide simulation results to support the fact that the bias is minimized at the daily frequency.

12.3 Empirical Implementation

12.3.1 The Data

The model is estimated using FTSE 100 returns from July 3, 1984 to December 29, 2006. Excluding weekends and holidays, we have 5686 daily observations. Summary statistics of the continuously compounded daily returns, in percentage terms, is provided in Table 12.1. Moreover, the time series graph of the return series is provided in Fig. 12.3.1. Table 12.1 shows that the unconditional distribution of the returns are negatively skewed and present excess kurtosis. The Jarque-Bera test rejects the null of normality at the 5% significance level. Moreover, the Ljung-Box test shows that the data exhibit autocorrelation.

Table 12.1 Summary statistics of FTSE returns

Statistic	FTSE 100
Mean	0.031
Volatility	1.035
Skewness	-0.557
Kurtosis	8.269
Min	-13.029
Max	7.597
Jarque-Bera	0.000
AR1	0.130
AR2	0.300
Num.Obs	5686

This table presents the descriptive statistics of FTSE 100 returns from July 3, 1984 to December 29, 2006. The Jarque-Bera test presents the p-value of the null hypothesis of normality. AR1 and AR2 show the p-values of the Ljung-Box test for autocorrelation of first and second order, respectively.

12.3.2 The Estimation Method

For the estimation method we use the Markov Chain Monte Carlo (MCMC). The MCMC is a numerical integration method that is able to deal with multidimensionality and nonlinearity, and can be used to estimate latent variables in complex models.

Let Y be the vector of observed stock returns, H the vector of state variables, and Θ the vector of parameters. The MCMC algorithm, based on the Clifford–Hammersley theorem, states that the joint distribution $p(\Theta, H|Y)$

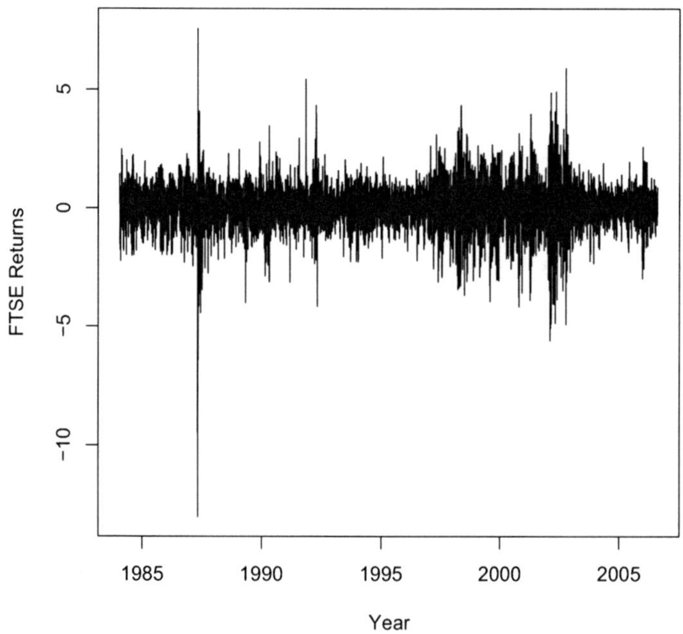

Fig. 12.1 FTSE returns.

can be characterized by its full conditional distributions $p(\Theta|H,Y)$ and $p(H|\Theta,Y)$. The algorithm therefore samples from $p(\Theta,H|Y)$ by sampling from the conditional distributions $p(\Theta|H,Y)$ and $p(H|\Theta,Y)$. Then, the Bayes rule factors the joint distribution into its components:

$$p(\Theta,H|Y) \propto p(Y|H,\Theta)p(H|\Theta)p(\Theta) \tag{12.4}$$

where $p(Y|H,\Theta)$ is the likelihood function, $p(H|\Theta)$ is the distribution of the state variables, and $p(\Theta)$ is the distribution of the parameters, also called the prior [11]. The conditional posterior of a parameter is derived from Equation (12.4) by ignoring all the terms that are constant with respect to that parameter.

The bivariate density function that we use is

$$f(B) = \frac{1}{2\pi|\Sigma|^{\frac{1}{2}}} \exp[-\frac{1}{2}(B-E(B))'\Sigma^{-1}(B-E(B))] \tag{12.5}$$

thus, the likelihood function is simply given by $\prod_{t=1}^{T} f(B)$, with

$$B = \begin{pmatrix} y_{t+1} \\ v_{t+1} \end{pmatrix}$$

$$E(B) = \begin{pmatrix} \mu + \xi^y_{t+1} J^y_{t+1} \\ \kappa(\theta - V_t) + \xi^v_{t+1} J^v_{t+1} \end{pmatrix}$$

$$\operatorname{cov}(y_{t+1}, v_{t+1}) = \Sigma = \begin{pmatrix} V_t & \rho\sigma_v V_t \\ \rho\sigma_v V_t & \sigma_v^2 V_t \end{pmatrix}$$

where $y_{t+1} = Y_{t+1} - Y_t$, $v_{t+1} = V_{t+1} - V_t$.

As explained previously, our joint distribution is given by the product of the likelihood times the distributions of the state variables times the priors of the parameters, more specifically:

$$\begin{aligned} \textit{Joint Density} = & \left[\prod_{t=1}^{T} f(H)\right] * \left[\prod_{t=1}^{T} \left(f\left(\xi^y_{t+1}\right) * f\left(\xi^v_{t+1}\right) * f\left(J^y_{t+1}\right) * f\left(J^v_{t+1}\right)\right)\right] \\ & * \left[f(\mu) * f(\kappa) * f(\theta) * f(\rho) * f\left(\sigma_v^2\right) * f(\mu_y) * f(\rho_J) * f\left(\sigma_y^2\right)\right] \\ & * \left[f(\mu_v) * f(\lambda_y) * f(\lambda_v)\right] \end{aligned} \tag{12.6}$$

The distributions of the state variables are given by: $\xi^v_{t+1} \sim \exp(\mu_v)$; $\xi^y_{t+1} \sim N(\mu_y + \rho_J \xi^v_{t+1},\ \sigma_y^2)$; $J^y_{t+1} \sim Bern(\lambda_y)$ and $J^v_{t+1} \sim Bern(\lambda_v)$. Following [8] we impose little information through our priors. They are as follows: $\mu \sim N(0,1)$, $\kappa \sim N(0,1)$, $\kappa\theta \sim N(0,1)$, $\rho \sim u(-1,1), \sigma_v^2 \sim IG(2.5, 0.1)$, $\mu_y \sim N(0,100)$, $\rho_J \sim N(0,4)$, $\sigma_y^2 \sim IG(5,20)$, $\mu_v \sim G(20,10)$, $\lambda_y \sim B(2,40)$ and $\lambda_v \sim B(2,40)$.

After the derivation of the conditional posteriors of the parameters and state variables, we implement MCMC by sampling directly from the conditional distributions when they are known in closed form. This case is called a Gibbs sampler. When the distributions are not known in closed form, a general approach called Metropolis–Hasting is used. The latter consists in sampling a candidate draw from a proposal density and then accepting or rejecting the candidate draw based on an acceptance criterion. We implement our MCMC by sampling iteratively from

$$\begin{aligned} \textit{Parameters}: &\ p(\Theta_i | \Theta_{-i}, J, \xi^y, \xi^v, V, Y), \quad i = 1, \dots, k \\ \textit{Jump times}: &\ p(J_t | \Theta, J_{-t}, \xi^y, \xi^v, V, Y), \quad t = 1, \dots, T \\ \textit{Jump sizes}: &\ p(\xi^y_t | \Theta, J, \xi^y_{-t}, \xi^v, V, Y), \quad t = 1, \dots, T \\ &\ p(\xi^v_t | \Theta, J, \xi^v_{-t}, \xi^y, V, Y), \quad t = 1, \dots, T \\ \textit{Volatility}: &\ p(V_t | \Theta, V_{t+1}, V_{t-1}, J, \xi^y_{-t}, \xi^v, Y), \quad t = 1, \dots, T \end{aligned}$$

12.3.3 The R Program

Our MCMC algorithm is implemented in R language. To the best of our knowledge, it is not yet possible to find in R a package dealing with Stochastic Volatility models with double jumps. In order to implement our MCMC procedure we wrote an algorithm that partially used functions in packages such as Rlab, MCMCpack, and msm.

The first thing that the algorithm requires is starting values that we first generate. The volatility vector was created using a three-month rolling window. We considered as jumps, differences in returns and in volatility that were above three standard deviations from the mean (after accounting for outliers). In a second step, we call a function that we programmed to implement either the Gibbs sampler or Metropolis–Hasting depending on whether or not we know the closed form of the parameters' distributions. For the jump sizes we have two cases: we sample from the conditional posteriors that we have derived when there are jumps, and from the prior distributions in the other case (no jumps) since the data does not provide further information in those cases.

The outputs of our function are the sampled parameters per iteration and the vectors of state variables. The mean of each parameter over the number of iterations gives us the parameter estimate. Then, we evaluate the convergence using trace plots which show the history of the chain for each parameter and are useful for diagnosing chains that get stuck in a region of the state space, and the ACF plots which are used to analyze the correlation structure of draws [11]. Finally, we check the performance by analyzing the residuals using the normal QQ plot and the Mean Squared Errors. We refer the reader to the CD accompanying this book for further specific details about the program.

12.3.4 The Results

We run the algorithm using 50,000 iterations with 5,000 burn-in iterations. Table 12.2 provides parameter posterior means and their computed standard errors. The return mean (μ) is close to the daily return mean from the data (0.0314). Note that due to the jump components, the long-term mean of volatility is given by $\theta + (\mu_v * \lambda_v)/\kappa$[1] which in our case equals 0.1795. We would expect this value to be close to the variance of returns 1.0353 but this is not the case. Numerically, we can explain this finding by analyzing θ, μ_v, and κ. As can be seen for Table 12.2, this smaller variance is due to the fact that the values of θ (0.1795) and μ_v (0.0008) are very small compared to what Eraker, Johannes and Polson [8] have found for the S&P500.[2] More-

[1] See [8].

[2] These authors found that θ and μ_v are equal to 0.5376 and 1.4832, respectively.

over, the values of κ in our case is rather big (0.9727) compared to 0.026 for the S&P500. These results may suggest that there should be other possible correlation structures between some of the parameters that need to be accounted for in the implementation of the MCMC algorithm. We have left this issue for further research.

Table 12.2 Estimation results

Parameter	Value
μ	0.0323 (0.0086)
μ_y	-0.0226 (0.1068)
μ_v	0.0008 (0.0004)
θ	0.1795 (0.0148)
κ	0.9727 (.03467)
ρ	-0.0506 (0.2594)
ρ_J	-0.0097 (2.0034)
λ_y	0.0477 (0.0652)
λ_v	0.0477 (0.0652)
σ_y	1.2164 (0.5327)
σ_v	0.3414 (4.0018)

This table presents the model's parameter estimates. Standard errors are in parentheses. μ represents the mean of returns, σ_y is the standard deviation of returns jump size. μ_y is part of the mean jump size in returns, and so is ρ_J as $\xi^y|\xi^v \sim N(\mu_y + \rho_J \xi^v, \sigma_y^2)$. μ_v represents the mean jump size in volatility, and θ is part of the long-term mean of volatility. σ_v is part of volatility of volatility, κ represents the speed of volatility mean reversion, ρ is the correlation coefficient between the error terms, and λ_y and λ_v are the jump intensity in returns and volatility, respectively.

On other hand, σ_y and σ_v are 1.2164 and 0.3414, respectively. These values are quite close to the values obtained by [8] for their Nasdaq case. Finally, the intensity of the jumps is higher than what we would expect. We got 0.0477 while from the data we would expect an intensity of around 0.02. However, this empirical intensity was computed considering that differences in returns above three standard deviations from the mean could be considered as jumps. Some caution must be taken since this empirical intensity is very sensitive to the number of standard deviations used as cutting point in defining the jumps.

One nice feature of the MCMC method is the possibility of estimating the latent variables. In our case the results of the estimated jump times, jump sizes in return and volatility, and the volatility paths are shown in Fig. 12.2. In the top-right panel the jump times are represented by jump probabilities meaning that a high probability can be associated with a high possibility of jump. We notice that the biggest probabilities of jumps are around 1987 and 2003 coherent with the financial crashes during those years. This is confirmed by the highest values of volatility that can be seen in the top-left panel, as well as the biggest jump sizes in returns and volatility in the lower panels.

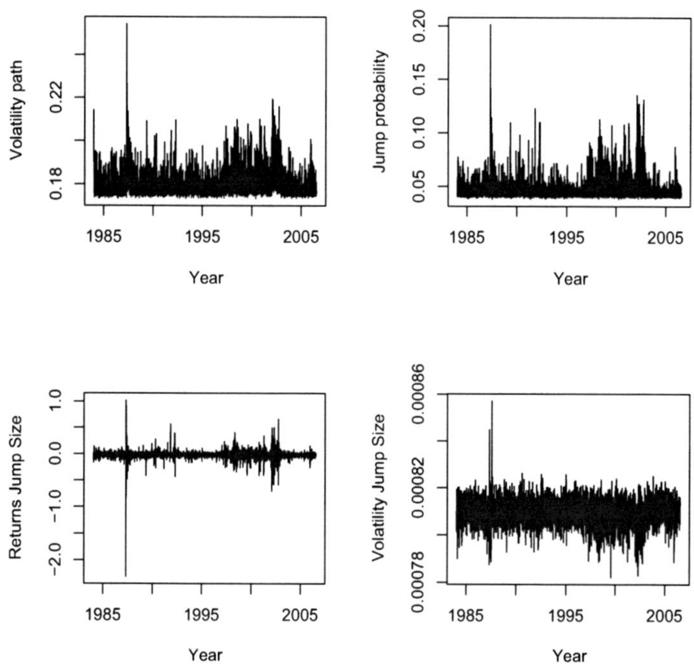

Fig. 12.2 In the top-left panel we represent the volatility path, in the top-right panel we have the estimated jump times. In the lower panels we represent the estimated jump sizes in returns on the left and the estimated jump size in volatility on the right.

The ability of the model to suit the data is accessed through the Mean Squared Errors and the normal QQ plot. The standardized error is

$$\frac{Y_{(t+1)\Delta} - Y_{t\Delta} - \mu\Delta - \xi^{y}_{(t+1)\Delta} J^{y}_{(t+1)\Delta}}{\sqrt{V_{t\Delta}\Delta}} = \varepsilon^{y}_{(t+1)\Delta} \tag{12.7}$$

Our Root Mean Squared Error value is 2.360819 which compared to the return mean is quite high. The QQ plot showed that the errors do not quite follow the normal distribution signaling some model misspecification.[3]

[3] Due to space limitation, additional information is available on request.

12.4 Conclusion and Future Venues of Research

In this chapter we present the implementation of the stochastic volatility model with jumps in return and volatility of [8] using an algorithm written in R language. Using Markov Chain Monte Carlo as the estimation method, our algorithm produces parameter estimates and state variable paths. The program is also able to conduct convergence and performance analysis on the output using trace plots, ACF plots, and analyzing the error terms.

For the empirical part of the chapter we use FTSE 100 daily returns from July 3, 1984 to December 29, 2006. Our results confirm that it is possible to estimate stochastic volatility models with double jumps in R using MCMC. However, more work needs to be done on the sampling of some of the parameters which were different from what we would expect, specifically θ which is part of the long-term mean of volatility, κ the speed of mean reversion, and μ_v the mean jump size in volatility. We suspect that there are possible correlation structures between some of these parameters that need to be accounted for in the implementation of the MCMC algorithm. We expect to contribute with this program to the existing library of programs that can be used either by academics or practitioners alike.

References

1. Bakshi, G., Cao, C., Chen, Z.: Empirical performance of alternative option pricing models. Journal of Finance **52**, 2003 – 2049 (1997)
2. Bates, D.: Jumps and stochastic volatility: Exchange rate processes implicit in Deutsche Mark options. Review of Financial Studies **9**, 69–107 (1996)
3. Bates, D.: Post-'87 Crash fears in S&P 500 futures options. Journal of Econometrics **94**, 181–238 (2000)
4. Black, F., Scholes, M.: The valuation of options and corporate liabilities. Journal of Political Economy **81**, 637–654 (1973)
5. Chernov, M., Gallant, A.R., Ghysels, E., Tauchen, G.E.: A New Class of Stochastic Volatility Models with Jumps: Theory and Estimation. SSRN eLibrary (1999). DOI 10.2139/ssrn.189628
6. Cox, J.C., Ross, S.A.: The valuation of options for alternative stochastic processes. Journal of Financial Economics **3**, 145–166 (1976)
7. Duffie, D., Pan, J., Singleton, K.: Transform analysis and asset pricing for affine jump-diffusions. Econometrica **68**, 1343–1376 (2000)
8. Eraker, B., Johannes, M., Polson, N.: The Impact of Jumps in Volatility and Returns. Journal of Finance **58**, 1269–1300 (2003)
9. Heston, S.: A closed-form solution for options with stochastic volatility with applications to bond and currency options. The Review of Financial Studies **6**, 327–343 (1993)
10. Hull, J.C., White, A.: The pricing of options on assets with stochastic volatilities. Journal of Finance **42**, 281–300 (1987)
11. Johannes, M., Polson, N.: Mcmc methods for continuous-time financial econometrics. In: Y. Ait-Sahalia, L. Hansen (eds.) Handbook of Financial Econometrics. Elsevier, New York (2003)

12. Merton, R.C.: Option Pricing when Underlying Stock Returns Are Discontinuous. Journal of Financial Economics **3**(1–2), 125–144 (1976)
13. Pan, J.: The jump-risk premia implicit in options: Evidence from an integrated time-series study. Journal of Financial Economics **63**, 3–50 (2002)
14. Rubinstein, M.: Nonparametric Tests of Alternative Option Pricing Models Using All Reported Trades and Quotes on the 30 Most Active CBOE Option Classes from August 23, 1976 through August 31, 1978. Journal of Finance **40**(2), 455–80 (1985)
15. Scott, L.: Pricing stock options in a jump-diffusion model with stochastic volatility and interest rates: Applications of Fourier inversion methods. Mathematical Finance **7**, 413–426 (1997)
16. Scott, L.O.: Option pricing when the variance changes randomly: Theory, estimation, and an application. Journal of Financial and Quantitative Analysis **22**, 419–438 (1987)
17. Vinod, H.D.: Hands-on Intermediate Econometrics Using R: Templates for Extending Dozens of Practical Examples. World Scientific, Hackensack, NJ (2008). URL `http://www.worldscibooks.com/economics/6895.html`. ISBN 10-981-281-885-5

Index

H.D. Vinod (ed.), *Advances in Social Science Research Using R*,
Lecture Notes in Statistics 196, DOI 10.1007/978-1-4419-1764-5,

Lightning Source UK Ltd.
Milton Keynes UK
15 September 2010

159864UK00003B/90/P

9 781441 917638